The Smartphone

ANATOMY OF AN INDUSTRY

Elizabeth Woyke

THE NEW PRESS

NEW YORK
LONDON

Requests for permission to reproduce selections from this book should be mailed to:
Permissions Department, The New Press, 120 Wall Street, 31st floor, New York, NY 10005.

Published in the United States by The New Press, New York, 2014
Distributed by Perseus Distribution

LIBRARY OF CONGRESS CATALOGING-IN-PUBLICATION DATA

Woyke, Elizabeth.
  The smartphone : anatomy of an industry / Elizabeth Woyke.
    pages cm
  Includes bibliographical references and index.
  ISBN 978-1-59558-963-7 (paperback)—ISBN 978-1-59558-968-2 (e-book) 1. Smart
phones. 2. Cell phones. 3. Telecommunication. 4. Electronic industries. I. Title.
  HE9713.W69 2014
  384.5'34—dc23   2014020781

The New Press publishes books that promote and enrich public discussion and
understanding of the issues vital to our democracy and to a more equitable world.
These books are made possible by the enthusiasm of our readers; the support of a
committed group of donors, large and small; the collaboration of our many partners
in the independent media and the not-for-profit sector; booksellers, who often hand-sell
New Press books; librarians; and above all by our authors.

www.thenewpress.com

Composition by dix!
This book was set in Minion

Printed in the United States of America

10 9 8 7 6 5 4 3 2 1

For my parents, John and Priscilla

Contents

1

From the Simon
to the BlackBerry

Martin Cooper with Motorola's Dyna-
Tac prototype in 1973 (*Martin Cooper*)

On April 3, 1973, Martin Cooper dialed his way into history. As the general manager of Motorola's systems division, he had flown to New York City to unveil a prototype of the world's first handheld cellphone. The 28-ounce phone, which had a long antenna, a thin body, and a protruding bottom "lip," making it resemble a boot, isn't sleek by current standards, but it was revolutionary. Until 1973 a mobile phone required so much power it had to be tethered to a car's electrical system or an attaché case containing a huge battery. The phone that Cooper and his team had developed—the DynaTAC—fit right in the palm of his hand.

Reporters had gathered at the Hilton hotel in Midtown Manhattan to see the new phone. Cooper was nervous. "There were thousands of parts in the thing; it was hard to keep it running," he recalls. "We had people at the hotel still working at midnight [the night before] to make sure the phone would be able to make calls."[1] Cooper got lucky. The phone functioned flawlessly that day, both before the press conference when he placed a call from the bustling street in front of a reporter, and later at the event, where he made a number of calls, even letting one young journalist dial her mother in Australia. "At that time, not everyone in the world thought people needed cellphones," he said, "but the reporters were quite enthusiastic." Today Cooper is universally acknowledged as the creator of the cellphone and the first person to make a cellphone call in public. His story gave the world a straightforward starting point for understanding cellphone history.

In contrast, there is no consensus on the smartphone's origins. A number of people think it was born in 2007, when Apple cofounder Steve Jobs proudly showed off the first iPhone at the Macworld conference in San Francisco. But what many people either forget or do not know is that phones with smartphone features had already been on sale for more than a decade.

Some experts believe smartphones emerged from cellphones when manufacturers began squeezing sophisticated programs and Web-browsing features into their handsets. Others say personal digital assistants (PDA), with their touchscreens and open operating systems, were the real progenitors of the smartphone. A third camp thinks pagers and messaging devices, including early BlackBerrys, paved the way by introducing mobile data and e-mail to a broad audience.

The question hangs on how you define a smartphone. Generally speaking, a smartphone distinguishes itself from a cellphone by running on an open operating system that can host applications (apps) written by outside developers. The apps expand the phone's functionality, giving it computerlike capabilities, and can be downloaded and installed by users, not just pre-installed by smartphone companies. Smartphones also have a number of built-in features that basic phones typically do not, including touchscreens that can sense multiple-finger swipes, high-definition displays, fully Internet-capable browsers, advanced software that automatically grabs new e-mails, and high-quality cameras, music, and video players.

It took more than a decade to cram all these features into one handheld device. The earliest smartphones came from IBM, Nokia, Ericsson, Palm, and Research In Motion/BlackBerry. Though these phones pushed boundaries in the 1990s and early 2000s, they were all limited in some way, especially in their Internet and app access. Most of these early smartphones were not sales hits. Some were famous flops. But all contributed to the smartphones we now carry in our pockets, whether they are iPhones, Android phones, Windows Phones, or BlackBerrys.

IBM AND THE SIMON

IBM's Simon phone was born of twin desires: a talented engineer's to tackle the challenge of creating a portable, wirelessly connected computing device and IBM's to burnish its image by unveiling never before seen, futuristic gadgets. Though the Simon was commercially available for less than a year, many regard it as the world's first smartphone.

Frank J. Canova Jr. was the gutsy engineer behind the Simon. Gary Wisgo, who was Canova's engineering manager at IBM, says Canova had two rare abilities. In an industry known for its narrow specializations, Canova was skilled at both hardware and software. He also had a knack for seeing the future. "A lot of engineers just sit and design circuits; a lot of programmers just sit and design programs," says Wisgo. "Frank would look out at the industry, see what other companies were doing, and say, 'We should do this.' "

The early 1990s were a fertile time for dreaming up portable, wireless gizmos. Advances in microprocessor chips and other components made it possible to shrink computers into mobile devices. PCMCIA cards—credit card–sized "personal computer memory cards" that could expand a computer's storage or functionality—were sparking new ideas about computing capabilities. Cellular service was expanding across the country. And carriers were planning network upgrades that would make it easier to send and receive data on mobile devices.

Anticipating a lucrative new market, technology companies started concocting next-generation computing products. Apple was stealthily developing its Newton PDA and had publicly confessed an interest in wireless devices. AT&T was funding a start-up to create the EO, a PDA-tablet hybrid with a built-in cellular modem. IBM, which had produced PCs for a decade, was also investigating computer miniaturization and wireless connectivity, in its research lab in Boca Raton, Florida.

Canova worked in that lab as a senior technical staff member on IBM's advanced technology team. In mid-1992, the 35-year-old was experimenting with touch-sensitive glass panels that snapped onto the front of computer monitors. He decided to try making a touchscreen version of a phone keypad, using a computer monitor, a glass touch-screen overlay, and Microsoft's Visual Basic computer programming language. Within a few days Canova had a touch-responsive mock-up of the keypad—and a plan. He started talking to IBM management about creating a touchscreen handheld device that would combine calling and computing features.

IBM was game. It had recently kicked off a corporate initiative called Vision 95, which aimed to predict what computers, electronics, and computing services would look like in 1995. As part of the initiative, Jerry Merckel, an IBM manager who specialized in strategic planning, commissioned sleek wood-and-plastic models of Canova's phone concept and a set of matching PCMCIA cards. Dubbing it "tomorrow's computer," he explained how the cards would transform the

Model PCMCIA cards used in internal IBM presentations to obtain funding for the Simon project. Inserting the cards would expand the phone's functionality by adding GPS, camera, TV, and radio capabilities. (*Jerry Merckel*)

phone into a camera, or a music player or game player. Intrigued, Paul Mugge, who was an IBM vice president and the general manager of the Boca Raton office, approved funding.

Canova set about designing a device that would be a phone first and a computer second. "We wanted using it to be natural, like a [landline] phone, where you just lift the handset and dial numbers," he explains. "Unlike a computer, you wouldn't have to boot it or configure it." The phone's touchscreen design supported that ethos. Because touchscreens are thinner than built-in keyboards, users would be able to operate the device with one hand, like a regular phone. While tinkering with the touchscreen, Canova came to another realization: the phone's menu should be arranged like a computer's, with small icons for launching applications.

When IBM management saw Canova's touchscreen demos, they decided to include the phone in the company's technology showcase at the upcoming 1992 Computer Dealers' Exhibition (COMDEX)—then the computer industry's largest trade show, which was held in Las Vegas each November. IBM's mainframe business was collapsing, because its corporate clients were shifting their computing to lower-cost personal computers, and the company wanted to portray at least a portion of the company as still vibrant.

The phone project became a five-person effort. Wisgo says he was the manager, but Canova was the visionary. Wisgo's team had just 14 weeks to transform Canova's lab work into a prototype for COMDEX. The phone could make calls using a radio module IBM had sourced from Motorola, but the team still had to design the circuit board and find manufacturers for the touchscreen and battery. "Half of the device was in my mind; the other half was in pieces, splayed on a bench," Canova remembers. Writing software to support the phone's various functions was the most demanding task. Besides making calls, the phone needed to be able to send and receive e-mails, store the user's calendar and address book, and host a calculator. Users also needed to be able to move fluidly between the different features by switching screens—an early version of mobile multitasking. The team worked 80-hour weeks to meet the deadline. "We lived at the lab day and night," says Wisgo. "It was one of the most trying periods in my life."

Canova, who was a new father, brought his infant son to the lab on weekends so he could squeeze in time with him.

Glitches with the phone's code continued right up to COMDEX. "People would ask, 'Should we pull it?' and I'd say, 'Keep going,'" Mugge remembers. "We didn't have a backup plan." The team pressed on. On the first day of the trade show, after successfully demonstrating the phone's calling and e-mail features, attendees mobbed the IBM booth to get a closer look at the device. That week, *USA Today* published two articles highlighting the phone. The first marveled at the speed with which IBM developed the device. The second featured a photo of Canova holding it and described IBM's COMDEX demo as "seemingly flawless."[2]

Encouraged by the media attention, IBM agreed to fund the creation of a real product, and soon after, the Atlanta-based carrier Bell-South expressed interest in carrying it. IBM had been calling it a "personal communicator," but BellSouth named the phone Simon,

IBM Simon (*Wikimedia*)

after the children's game Simon Says and the electronic memory game that was popular in the late 1970s and 1980s. According to *Bloomberg Businessweek*, BellSouth's marketing managers thought the name "evoked simplicity" and would be easy for consumers to remember.[3]

Targeting an April 1994 release, IBM's Simon team swelled from 5 people to more than 30. Taking the phone from a prototype to a real product was challenging. Motorola refused to supply the phone's radios, explaining it would be "counterstrategic" for them to help IBM build a commercial smartphone, according to Wisgo. IBM got a replacement part from Mitsubishi, the Japanese electronics manufacturer, but the swap delayed the entire project.

The Simon team also wrestled with the phone's size and battery. The brick-shaped device ended up measuring 8 inches by 2.5 inches by 1.5 inches and weighing 18 ounces. "It was clunky, and we knew it, but we couldn't get it any smaller at the time," says Canova. Battery life was the single toughest issue. It topped out at an hour of talk time or eight hours of standby time, so IBM provided BellSouth with a thicker, double-capacity battery to sell separately. "We could have put a huge battery on [the Simon], but no one would have bought it," explains Wisgo. "It was too big already."

The Simon launched four months later than planned, at a price of $899. It boasted an advanced set of features. It supported multiple forms of communication: calls, e-mail, faxes, and pager messages. It had a "graphical user interface," meaning it was image- and icon-based. To navigate between phone, fax, and calendar functions, users simply touched the appropriate icons on the touchscreen, and they could customize their phones' features and functions by inserting different PCMCIA cards, similar to the way apps expand smartphone functionality today. The phone even included password protection, a note pad feature that saved typed memos, a world clock, and a puzzle-piece game called Scrabble®.

These features have since become commonplace in smartphones, but in the early 1990s they were unusual and, to some people, intimidating. BellSouth tried to counteract confusion by advertising the Simon as "Mobile Communications Made Simple,"[4] but news articles described the phone as everything from a portable computer to a "processor-based telephone"[5] to an "enhanced telephone and messaging unit."[6] Reviews tended to commend the phone's advanced features without advising people to buy it. *PC Magazine* said the Simon was "an impressive feat of integration" but cautioned that "while Simon is definitely much more than a phone, it's still not quite a PC."[7]

BellSouth sold fewer than 50,000 units of the Simon before the phone was discontinued in 1995. IBM, still in financial trouble, had already begun downsizing the Simon team. "It was a time of enormous restructuring," explains Mugge. "There was just no room or time for these forays into the future." Canova left IBM in 1994 when the company started closing the Boca Raton lab and moving projects to other locations. By then Canova had designed a successor to the Simon— the Neon, which was much smaller and had a unique "tilt sensor" that would rotate the screen when the phone was turned sideways. IBM tried to continue developing the Neon, but the project floundered due to limited resources and loss of talent.

The Simon was an innovation doomed by bad timing. It missed IBM's golden age of the 1980s. It also employed first-generation (1G) analog technology right before the United States moved to faster 2G digital networks. While analog cellphone networks were designed to support only voice communications, 2G networks enabled greater flexibility and efficiency in terms of mobile services and bandwidth usage, allowing carriers to serve more subscribers and cellphones to last longer on a single charge, among other benefits. The Simon was the first e-mail-capable phone, but sending and receiving e-mails on a 1G network could be tedious, as users had to dial in to a remote computer via a built-in modem, and then wait to download and upload messages. "Carriers hadn't gotten into digital networks yet, so we were trying to push data over an analog network, which is really never ideal," says Canova.

Popular Science pointed out the problem: "Like most cellular phones, Simon is an analog communications device that's subject to interference problems and spotty coverage outside urban areas. Digital cellular phones and services . . . are much better suited for sending messages and faxes. . . . For all of its futuristic touches, Simon is still firmly rooted in the limitations of today."[8]

Another cutting-edge feature the Simon didn't get to properly show off: its ability to support apps. Mugge believed the Simon would inspire developers to create PCMCIA cards "with all kinds of interesting functions" just like IBM's PCs had "attracted the minds and talents of people all over the world." But BellSouth and IBM didn't establish a formal Simon app development program. "Anybody who wanted to make an app had to jump through more hoops than common people would be willing to do," acknowledges Canova. Only one company outside IBM made the effort: an Atlanta-based software firm called PDA Dimensions. The app, called DispatchIt, was intended as a way for field service teams to keep a log of their work. It was useful but extremely expensive: it cost $299, and the corresponding PC software cost $2,999.[9]

The next smartphones would be 2G phones. With their speedier digital network connections, smaller dimensions, and more advanced features, these phones would fare better than the ill-fated Simon. And it was in Europe that they would first emerge.

FROM THE UNITED STATES TO EUROPE

Change is a constant in smartphone history. A company and geographic region will lead innovation for some years, but a challenger always emerges to overtake them.

In the pre-smartphone era the United States spearheaded the development of cellphone technology. Bell Labs, AT&T's former R&D division, started formulating the principles of cellular networks in the 1940s and continued through the 1970s. Motorola, under Martin Cooper, kept the United States at the forefront of the technology in the 1970s by conceiving the first portable cellphone. But the United States lost momentum because the Federal Communications Commission (FCC) took nearly a decade to decide which companies would get licenses to offer commercial cellular networks. AT&T was lobbying for the privilege, but so were Motorola and its corporate partners. "The question was, 'Is the government going to consider cellular to be part of the phone system, which was an AT&T monopoly at the time, and give [the radio frequency] to AT&T or try to do something different, which

Motorola and its customers were urging?' " says Sheldon Hochheiser, the institutional historian at the IEEE History Center, a nonprofit that preserves and promotes electrical engineering and computing history. Motorola and AT&T's rivalry was, in fact, the catalyst for the Dyna-TAC and Cooper's now famous April 1973 press conference. "Back then, AT&T had enormous power and Motorola was a little company; we had to build up some credibility," explains Cooper.

In 1974, the issue moved further into the political arena when the Department of Justice filed an antitrust lawsuit against AT&T that eventually forced the company to break up its Bell system. In the end, the FCC decided that each market, which basically meant each metropolitan area, should have two cellular licenses—one for an incumbent landline carrier, such as AT&T (which passed its licenses to its regional subsidiaries after its 1984 breakup), and one for a company, such as Motorola, that wasn't a traditional carrier. "It was strictly politics," recalls Richard Frenkiel, a former Bell Labs employee who helped pioneer cellular-system engineering with Joel Engel, another Bell Labs engineer.[10] "We had a working system [in Chicago] in 1978, but it didn't go commercial until 1983, because people were arguing about an AT&T monopoly."

Frenkiel has described that Chicago system as the world's "first true cellular system" and capable of "huge capacity,"[11] but while the FCC deliberated, northern Europe pulled ahead. In 1981, Norway and Sweden activated their first cellphone networks. Denmark and Finland followed in 1982. Cellphones were an ideal way for Scandinavian countries to connect their far-flung residents across long distances and heavy snow, because cellular networks are cheaper to build than landline phone networks. The disadvantage was that these early cellular networks were often incompatible with each other. "Most European countries had slight variations in their 1G networks," says Nigel Linge, a telecommunications professor and historian at the University of Salford in Britain. "If you had a mobile phone in England in the 1980s, it stopped working at the English Channel." Since Europeans frequently cross national borders, they quickly realized they needed a pan-European cellular technology.

A mobile telecommunications standard outlines the technical way cellphones interact with networks so subscribers' calls and data can be processed and transmitted, and in 1987, 13 European countries, including Britain, France, Germany, and Italy, agreed to adopt a standard called GSM, which has come to mean Global System for Mobile communications, for their 2G networks. Europe then deftly navigated the transition from 1G to 2G networks, launching 2G/GSM ones in 1991, earlier than any other region of the world. The rest soon followed, as GSM became the de facto global standard, and by 1997, 108 countries across the world had commercial GSM networks up and running.

In contrast, U.S. carriers didn't deploy 2G until 1995, and U.S. regulators took a free market approach, letting each carrier decide which 2G standard it would use. Some American carriers chose GSM, but Sprint, and the carriers that later merged to become Verizon Wireless, chose a competing standard called code division multiple access (CDMA)— a newer technology than GSM that offered carriers greater capacity, so they could serve more subscribers. The arrival of 2G unified the European wireless industry but fragmented that of the United States.

European standardization helped Sweden's Ericsson and Finland's Nokia grab early leads over Motorola in 2G cellphones and smartphones. GSM enabled European phone makers to sell in many countries without major modifications, and the prospect of international high-volume sales was a great incentive to innovate. Advanced networks also made device innovation easier in Europe, since European carriers were eager to offer sophisticated phones that would highlight their networks' benefits.

By the mid-1990s, nearly one out of every three Finns and Swedes owned cellphones, which was more than twice the rate for Americans and the rest of Europe. In 1997, the *New York Times* declared Finland the "most wired nation in the world," citing its enthusiastic adoption of cellphones, the Internet, and online services, and attributed Finland's leadership in communications technology to its "high educational levels, the government's spending in basic research, and the long winter nights." [12] Both Finland and Sweden also had competitive wireless markets (which reduced cellular service prices for consumers), strong engineering traditions, and a focus on economic growth through exports, due to their relatively small populations.

Nokia and Ericsson

Jari Kiuru joined Nokia in 1995 without knowing what he would be doing. Nokia would only say it was "building something unique that no one had seen before." The project was based in Tampere, an inland city several hours from Nokia's main offices in Helsinki. Almost everyone on the team was in their early to midtwenties. "Nobody was selected because they had prior experience," says Kiuru. The team was charged with creating a phone-PDA hybrid, something that Nokia had been thinking about since the early 1990s. Kiuru initially served as a product manager, but within a few months he was promoted to program manager, which meant he oversaw the project's entire development, from R&D to manufacturing to marketing. The group's first product was the Nokia 9000 Communicator—a 6.8-inch-long, 14-ounce éclair-shaped device that looked like a chubby cellphone when closed but opened to reveal a PDA, with a large, 4.5-inch screen and a built-in QWERTY keyboard that had letter keys laid out in the same configuration as a standard English-language computer keyboard.

Nokia Communicator 9000 (*textlad/Flickr*)

Nokia didn't consider the 9000 Communicator to be a smartphone—a term that was already in circulation, although its meaning varied. Companies, including AT&T, had first used the term in the 1980s to describe landline phones that had special features. Those noncellular smartphones included voice-controlled office phones that could dial preset numbers when commanded and home phones that had built-in modems and touchscreens so users could check their bank account balances, transfer funds, and pay bills. When manufacturers began giving cellphones handheld computer features, people started calling them smartphones, too. So when Nokia dubbed the 9000 a "communicator," it was attempting to create a new product category that would rank it above smartphones in price and functionality. "A smartphone was just a phone with advanced functionality," says Kiuru. "Communicator sounded like a miniature computer that was advanced in communications." It also had the futuristic ring of the voice communication devices used on *Star Trek*.

In many respects the Communicator was a refined, more powerful version of the Simon. It weighed 0.9 pounds and was almost seven inches long, which made it slightly lighter and shorter than the Simon. Both phones could support apps, send faxes and e-mails, and store information, such as contact lists, notes, and calendars. The Communicator could also do things the Simon couldn't, such as browse the Web and run third-party apps built with a software development kit (SDK). Kiuru says the latter feature qualifies the Communicator as the "first genuine advanced smartphone," since releasing an SDK to help outsiders write apps would later become standard practice.

The Communicator's myriad features and functions helped it stand out but also made it much more difficult to produce. While a typical cellphone at that time had 25 to 30 component suppliers, the Communicator had 250. And while a typical cellphone had 100 to 150 discrete components, the Communicator had 1,080. To make the phone as compact as possible, Nokia devised a seven-layer circuit board, which solved another problem: some smartphone chips emit signals and can interfere with each other if placed too close together. The multiple-layer circuit board separated the chips with layers of a nonconductive material, such as plastic, instead of large amounts of space.[13] But the board was so complicated to build that a fabrication mix-up brought the entire Communicator project to a halt less than two months before its scheduled launch. "In early July 1996, everything stopped working," recalls Kiuru. "You couldn't even make a phone call."

It was a setback Nokia couldn't afford. To drum up press and public interest in the Communicator Nokia had announced the phone's on-sale date and time (August 15, 1996, at noon) five months in advance. Publications including *The Times* of London had mentioned it in articles,[14] and Nokia management was intent on meeting its highly publicized deadline. Kiuru's team started retracing their steps to locate their mistake. Late one night a sharp-eyed radio design engineer pinpointed the problem: the manufacturer Nokia had hired to produce the circuit board had flipped one of the seven layers upside down by mistake.

The Communicator team recovered from that episode and delivered the phone in time for an unusually flashy, London-based launch that included a public drawing for a deeply discounted Communicator and paid appearances by British athletes who had won medals at the 1996 Olympic Games. Reviewers marveled at the device's features even while balking at the $1,500 price tag and its size. *Barron's* highlighted its "seriously high gee-whiz quotient,"[15] while Reuters noted, "The Communicator puts more power in the palm of your hand than most business computers put on your desktop just a few years ago."[16] *The Times* wrote, "It may not be particularly pretty, but it is almost certainly the most advanced mobile phone in the world."[17]

In 1997, Nokia decided to bring a variant of the 9000 Communicator to the United States—the 9000i Communicator, which had updated software and could operate on American GSM networks. Six months later the sleeker and lighter 9000il Communicator debuted and became Nokia's new flagship U.S. device. Even with a redesign, the 9000il was an elaborate-looking GSM phone, which made it a hard sell in the United States, where GSM was yet to be established nationwide. It didn't help either that Americans were also still wedded to their pagers. Some U.S. carriers did buy the phone, but only in small volumes, so they could test it among their high-end business customers. To spur business interest in the 9000il, Nokia ran ads in the *Wall Street Journal* touting the phone's corporate features and portability.

They informed readers that the 9000il would let them "update [their] stock portfolios over lunch" and send a fax while sipping their "morning latte grande." [18]

The Communicator met with more success in Europe. In late August 1996, Nokia's then CEO, Jorma Ollila, said the company was "shipping [all the Communicators] we can make," [19] which amounted to tens of thousands of units a week, to European carriers.

Nokia's other Communicator-related accomplishment was that it released a smartphone before its main rivals, Ericsson and Motorola. "We knew Ericsson was trying to develop something similar," says Kiuru, and he was right. Nils Rydbeck, the chief technology officer of Ericsson's mobile phone division from 1985 to 2001, says Ericsson started thinking about smartphones in 1995 and began developing them in 1997. Ericsson called its smartphone strategy "the melting pot," which basically meant future phones would be amalgams of several gadgets, such as cameras, computers, and music players. "All these surrounding devices would go into the mixture and come out in a mobile phone as something new," says Rydbeck.

Ericsson launched its melting pot smartphone, the R380, in 2000. It improved on the original Communicator in a few ways. It was much cheaper, with a price of around $600. It was also much smaller—more than 1.5 inches shorter and less than half the weight of the 9000 Communicator. Instead of a built-in QWERTY keyboard, the R380 had a narrow, 12-button keypad that flipped open to reveal a touchscreen and a full, virtual keyboard. When held horizontally, the R380 screen measured 3.3 inches long and 1.3 inches high, giving users adequate room to view Web pages or the phone's address book, calendar, and note pad features. "The idea was to introduce the idea of a smartphone to people who were used to normal phones," says Rydbeck.

Kevin Lloyd, a former Ericsson global product manager who helped sell the R380 to carriers, thinks the phone was released too early. "It was nicely designed and had tremendous potential, but it was this kind of 'almost product,' " he recalls. "Companies were always asking for just one more iteration of the software, one more set of capabilities, before they would adopt it." Rydbeck acknowledges the phone could be "clumsy" but says Ericsson primarily wanted to test the smartphone market. He estimates the company sold at most "a couple hundred thousand" R380s. Nevertheless, the *Guardian* called it "a masterpiece of miniaturization" [20] while *Popular Science* named it one of the year's top 100 products. [21]

[Top] Ericsson R380s (*Tor Björn Minde/Ericsson*)
[Bottom] The Ericsson R380s was the first smartphone to have retail packaging that said "Smartphone" on it. (*Alex Bucher/www.oldericsson.ch*)

The R380 set a number of important smartphone precedents. It was one of the first post-Simon phones with a touchscreen. It was the first phone to run a very early version of the mobile-specific operating system Symbian, which would become a smartphone staple in the following decade. The R380 was also one of the first phones to use a technical standard called wireless application protocol (WAP) that streamlined Web pages for faster loading and easier viewing on phones.

The R380 was also the first smartphone to be explicitly marketed as a smartphone. Ericsson initially used "Smart Phone" to describe its GS88 phone, which looked similar to a Nokia 9000 Communicator. Ericsson created a GS88 prototype in 1997 and even designed its retail packaging—a box that said Smart Phone on its cover—but the company never publicly released the phone. "Weight was a little bit of an issue, and maybe the battery was not so great," recalls Tor Björn Minde, a longtime Ericsson employee who now serves as the company's head of research. "So it was probably decided to make [the GS88] smaller, to be the R380," which then became the first smartphone-branded phone. Says Rydbeck, "It was a good name for a new kind of phone that was more than a phone, that was also a camera, a music player, a Web browser, and a gaming machine."

PALM AND HANDSPRING

By the time mobile computing pioneer Jeff Hawkins established Palm Computing in 1992, he had several years of experience behind him: he had worked first at the chip maker Intel and then at a start-up called GRiD Systems, which made some of the world's first laptops and tablet computers. When he founded Palm Computing, he had a mission: billions of people would eventually carry mobile computers in their pocket, and he wanted Palm to build them.

After some missteps, Palm took a big stride toward that goal in 1996 when it launched the Pilot 1000 and 5000—the first PDAs Palm designed on its own rather than with a consortium of companies. The Pilot 5000 supported five times as much data and cost more than the Pilot 1000 ($369 versus $299), but both were 4.7-inch-long devices that let users save and access addresses, appointments, to-do lists, and memos, and to synchronize that information with that on their computers. The screens also had a touch-sensitive panel people could use to input data via a penlike stylus and a handwriting recognition system called Graffiti, which used simplified letter shapes that Hawkins had invented a few years earlier.

The Pilots won acclaim for being smaller, lighter, easier to use, and cheaper than previous PDAs, but Hawkins soon recognized that future PDAs would also need to include phone functionality. "Some time in 1997 or 1998, it dawned on me that cellphones and handheld computers were on a collision course," he explains. "PDAs and smartphones were all just a continuum of stuff in the same long-term mission of mobile computing."

By this time Palm had been acquired by U.S. Robotics, which was in turn acquired by networking equipment maker 3Com. The relationship between 3Com and Palm was awkward from the beginning, and so after 3Com refused to spin out Palm as a separate entity, Hawkins departed, along with Palm's president and CEO, Donna Dubinsky. Palm's head of marketing, Ed Colligan, soon joined them, and the trio established a company called Handspring in Mountain View, California, which produced a series of PDAs called Visor. The Visors ran Palm's operating system (Palm OS), which Handspring licensed from Palm.

Hawkins knew Handspring would need to offer smartphones to remain competitive. He also knew the start-up couldn't afford to produce a risky, complicated product. Handspring's solution was to include microphones inside its Visor PDAs and expansion slots in their backs. Sliding a cartridge called a "Springboard" into the slot would imbue the PDA with additional functionality, including phone calls.

In September 1999, Handspring launched its first Visors, at prices ranging from $149 to $249. About a year later, Visors became phones when Handspring introduced a $299 VisorPhone cartridge it built with the help of the Belgian radio technology company Option International. The module included a speaker, an earpiece, and an antenna that allowed users to make and receive calls. Users could also check e-mail and browse the Web, because the VisorPhone doubled as a wireless modem.

Reviews were mixed. Walt Mossberg, the *Wall Street Journal*'s former gadget reviewer, called the VisorPhone "a clever gizmo" and "a good choice for cutting down the clutter"[22] of carrying both a cellphone and a PDA. Ed Baig, the *USA Today* gadget columnist, described it as a "hybrid telephone solution that, while far from perfect, is the best I have seen."[23] But one *New York Times* article said, "It's difficult to imagine most self-respecting executives talking into their organiz-

Handspring VisorPhone (*Waldohreule/ Wikimedia*)

ers," adding that the semicircular VisorPhone cartridge made the PDA "look as if it's wearing a beanie." [24] Handspring sold fewer than 50,000 VisorPhones, even after slashing the module's price to $49 in 2001. Nonetheless, it gained experience in radios and wireless data, and Handspring's next big leap would be to real smartphones.

BLACKBERRY

In 2002, a smartphone appeared with an entirely different lineage. It was the BlackBerry 5810, a wireless handheld-plus-phone made by Research In Motion (RIM).

RIM was founded by Mike Lazaridis and Doug Fregin in 1984. The two had been friends since grade school in Windsor, Canada, united by their zeal for tinkering with gadgets. In high school, they built a basic computer together and found a mentor in their technical-shop teacher, who advised them not to get "too hooked" on computers because "Someday the person who puts wireless and computers together is really going to make something."[25] Both Lazaridis and Fregin went on to become engineering students, but Lazaridis dropped out of the University of Waterloo to develop the business that would become RIM and convinced Fregin to join him as cofounder. Lazaridis never forgot the advice of his high-school teacher, and by 1987 he had turned his attention to the emerging field of wireless data. After spending years building wireless modems for handheld devices and laptop computers, RIM launched its first pager in 1996. It was the world's first pocket-sized, two-way messaging pager and it allowed users to both send and receive messages wirelessly. Two years later, RIM followed with a smaller, lighter wireless e-mail device that had a screen capable of displaying eight lines of text, a trackwheel for navigating the pager's menu, and a QWERTY keyboard designed for "thumb typing."

Journalists and consumers often referred to the device as a PDA because it also had organizer features, such as a calendar, address book, and to-do list. RIM called it the RIM Inter@ctive Pager 950, and later the RIM 950 Wireless Handheld, but soon adopted a catchier name for its mobile devices: BlackBerry. The name was the creation of the California marketing firm Lexicon Branding that had also named Intel's Pentium line of microprocessors and General Motors's OnStar in-car communications system. Lexicon believed the name should be "colorful and connotative rather than dull and descriptive"[26] and thought RIM's devices, with their small, slanted, oval buttons, resembled a strawberry. But since "straw" is a slow-sounding syllable, and Lexicon's research had found that the letter B implied "reliability" to consumers, the firm recommended RIM use BlackBerry instead.[27]

As RIM grew, Fregin took less of a leading role than Lazaridis, who from 1992 shared CEO duties with the more business-minded Jim Balsillie. Like Hawkins, Lazaridis anticipated that wireless handhelds/PDAs and cellphones would converge, and in the late 1990s, he began talking to RIM employees about gaining a seat at the "MENS table," meaning Motorola, Ericsson, Nokia, and Siemens.* Needing some-thing big to break into the phone market, RIM decided to make the first North American PDA that could support both GSM and a recently in-troduced data technology called "general packet radio service" (GPRS).

First introduced in 2000, GPRS is a technology that carriers laid on top of their GSM networks to enable faster data rates. Dubbed 2.5G, GPRS divides data files, such as e-mails, into smaller digital units called "packets" that can be transmitted more efficiently, letting carriers charge subscribers for the amount of data they use rather than for the amount of time they spend online. The setup enabled GPRS

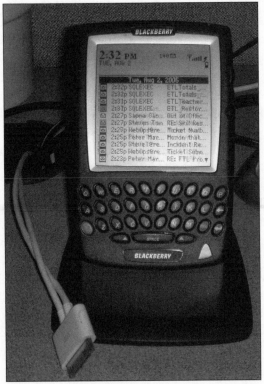

BlackBerry 5810 (*Steven Tom/Flickr*)

---

*Siemens, a German electronics and engineering company, produced cellphones from 1985 to 2005.

users to stay online all the time, saving them the hassle of dialing up an Internet connection when they wanted to send an e-mail or check for new mail. RIM's previous messaging devices had a similar "always on" capability, but that was because they operated on paging networks built to quickly deliver wireless messages. GPRS made it possible to offer that experience on a cellular network.

Once RIM set the goal of being first, it raced to release the 5810. To speed development it used the same plastic exterior casing as an earlier BlackBerry, so the 5810 resembled a PDA—a thin, 4.6-inch-long slab with a large, monochrome screen and a QWERTY keyboard. RIM also created different phone models for the United States and Europe because figuring out how to make one phone that could function in both regions would have taken longer. (The European version of the 5810 was called the 5820 and had different radios compatible with European GSM/GPRS networks.) RIM took another shortcut by not integrating a microphone or speaker into the 5810. Instead, it shipped the device with an in-ear headset, which users had to plug into the 5810 to make or answer calls. "It was a time-to-market trade-off made versus getting it absolutely right," says a RIM insider who worked on the project. "We were splashing into the market as the first and not the best, initially. We would build up to the best."

RIM promoted the $549 device as a "breakthrough in wireless convergence" that "lets you manage all of your business communications and information from a single, integrated wireless handheld."[28] Reviewers were less laudatory. Though *eWEEK* noted that the device "heeds [people's] call for fewer gadgets"[29] by enabling users to carry one device instead of two, *PC Magazine* called the 5810 a "not-so-convenient combo communicator"[30] and *Forbes* magazine said answering calls caused a "paroxysm of fumbling"[31] to get the earbud/mike combo in place.

The phone had other unexpected issues. European consumers considered the belt holsters RIM shipped with the 5820 to be deeply unstylish and refused to wear them. Without the protective holster, dust could collect in the device's earphone jack, which could affect audio quality. "That was a use-case that we just didn't put through its paces [before launch]," admits the RIM alum.

RIM was unfazed. The BlackBerry 5810/5820 accomplished the

important task of introducing RIM to carriers as business partners. Approximately 20 carriers across North America, Europe, and Asia agreed to sell the phone. One year later, RIM launched the BlackBerry 6210, which had a built-in microphone and speaker, and by then 50 carriers supported BlackBerrys.

The BlackBerry's seemingly instantaneous ability to receive and send e-mail provoked a level of user devotion that no other cellphone or smartphone had managed to inspire. In 1999 Lazaridis told a Reuters reporter, "People describe it as addictive. I'm addicted to this thing. Everyone in the company's addicted to this thing."[32] By 2000, major news outlets, including the business news channel CNBC, *Forbes*, and Canada's *Globe and Mail* newspaper, were using the term "Crackberry." The following year, *USA Today* took the Crackberry phenomenon mainstream with a long article entitled "BlackBerry: The 'heroin of mobile computing.' "[33]

In 2009 and part of 2010, at RIM's peak, the company controlled 20 percent of the global smartphone market, more than 50 percent of the U.S. smartphone market, had the top-selling smartphone brand in the United States, and shipped as many as 15 million phones per quarter. Though RIM's market share started sliding in late 2010—and hasn't stopped—BlackBerrys made their mark. Today all smartphone users expect their phones to have preinstalled e-mail "clients," or programs that fetch, send, and manage their messages.

RIM also built the first instant-message chat service for smartphones, BlackBerry Messenger (BBM). Today smartphone messaging apps are big business, with the most popular ones generating hundreds of millions of dollars a year in revenues. Though ones such as WhatsApp, which Facebook acquired in 2014, have many more features than the original BBM, they began with or quickly adopted the same basic idea: to be an always-on messaging service that lets users send unlimited messages for free or at a very low cost.

THE TREO

RIM had good reason to speed the 5810 to market. In 2002, Handspring began selling a smartphone line called Treo that would eventually replace its Visor series. Treos consolidated three gadgets—a phone, a wireless e-mail/messaging device, and a PDA—into a single device. As Hawkins recalls, "RIM was a potential competitor to us. We decided not to compete with them on the pager front but to go right to phones from PDAs. We knew RIM was going to end up in the phone space eventually too, [and the smartphone market was] the market that mattered."

Handspring announced its first Treo—the Treo 180—in October 2001, several months before the public even knew about the BlackBerry 5810. But since RIM rushed the 5810's release, the two phones went on sale around the same time, in early 2002.* Handspring and RIM, and their phones, had two major differences. Handspring was more focused on small and medium businesses and "mobile professionals"[34] who bought their own phones. RIM primarily sold BlackBerrys to financial services firms, corporations, and government agencies. These enterprises purchased not only devices: they also paid a per-user fee for

Handspring Treo 180 (*John Markos O'Neill/Flickr*)

*Handspring decided to ship the Treo 180 as a GSM device and add GPRS functionality later, through a software upgrade. That decision gave RIM leeway to market the 5810 as North America's first GSM/GPRS wireless handheld.

RIM's BlackBerry enterprise server software system, which enabled secure delivery of e-mails to their employees' BlackBerrys.

Unlike the BlackBerry, the Treo 180 looked like a phone—a big flip phone. While it had a large, monochrome screen like a PDA, the Treo 180 also sported a stubby antenna and a hinged lid that flipped up to serve as an earpiece for phone calls. At 4.3 inches by 2.7 inches by 0.7 inches, the Treo 180 was about the size of a sardine can; it was larger than a regular cellphone but shorter than a PDA.

USA Today's Ed Baig described the Treo 180 as "pretty slick" and "cool-looking."[35] The *Washington Post* said it was "surprisingly elegant."[36] But the design was the result of much compromise between Handspring and the carriers. Selling smartphones in significant numbers in the United States and many other countries required the support of the carriers, which dominated the smartphone retail market and subsidized device prices to make them more attractive to consumers. This system gave carriers substantial authority over cellphone and smartphone design.

Prior to the Treo, Handspring had sold the VisorPhone on its own website, but it wanted carriers to stock the Treo like they did other phones. Those agreements required negotiations that Hawkins found frustrating. "All of a sudden I couldn't build products I wanted to build," he says. "I had to build the products the carriers wanted to sell." Hawkins initially wanted Treos to have large touchscreens, like his PDAs: "My original thought was, these things shouldn't have keys. It should be all display, because you want to see information on it." Carriers disagreed, and they reminded Hawkins that cellphones had always had keypads, and consumers were used to them.

Carriers preferred the traditional, 12-key keypads that cellphones had used since the 1980s. Hawkins decided to put a BlackBerry-style keyboard on the Treo 180 instead. RIM's early BlackBerry keyboards were too wide to fit on a phone, so Handspring gave its version smaller keys and positioned the keys closer together. "RIM gets credit for inventing the thumb keyboard," says Hawkins. "We improved on it." But in September 2002 RIM sued Handspring for patent infringement. About six weeks later Handspring agreed to license RIM's keyboard patents for an undisclosed amount of money.

Keyboard issues were just one of the challenges the Treo 180 presented Handspring. Besides negotiating with carriers, Handspring had to strike licensing deals with network infrastructure and device makers that owned important wireless technology patents, learn about cellphone antennae, hire manufacturers to provide the phone's chipset and radio, and navigate cellphone radiation regulations. Hawkins says the stresses of producing the 180 almost killed the start-up. "It was so hard, and there was so much cost involved, and there were so many obstacles to overcome, it felt like it could put us out of business," he recalls.

People liked the Treo 180 for its relative compactness, and at $399 it was also more affordable than rival devices. But the phone's reviews were better than its sales, which, according to Dubinsky's records, were about 59,000 units. "It was not a failure, but it wasn't a rollicking success, either," says Hawkins. Handspring would keep refining the Treo concept, giving later Treos color screens, ditching the hinged lid, and shrinking the phone's overall size. In 2003, these efforts would come together in the Treo 600, a narrower, bar-shaped phone that tech publications such as CNET have named one of the most iconic smartphones ever.[37]

The Treo's other smartphone legacy was its open platform and strong developer relationships. Since Hawkins viewed the Treo (and the Visor and Palm's Pilots before it) as a minicomputer, he gave his devices open operating systems (Palm OS) and released free SDKs to outside developers, so they could build apps. "Desktop computers had thriving developer communities," he points out. "That was my model." So when the Treo 180 launched, it was compatible with more than 11,000 apps. Though not as technologically sophisticated as modern smartphone apps, those of the Palm OS addressed many of the same user needs as today's, such as editing photos, organizing grocery lists, and providing maps of the New York City subway system. At the time Palm had more than 175,000 developers worldwide, ranging from high school students to companies, and Handspring had more than 10,000 of its own registered developers. Treos were far ahead of BlackBerrys in this respect. RIM never became a developer favorite, and the BlackBerry's lack of compelling apps would later limit its consumer appeal.

SYMBIAN AND WINDOWS MOBILE

In 1998, a new platform entered the arena to compete with the Black-Berrys and Treos—Symbian, which Ericsson, Motorola, and Nokia purchased from British mobile computing company Psion. After establishing itself in the early 1980s as a maker of games and other software for computers, Psion started making handheld electronic organizers, which it also called "palmtop computers," in 1984. In contrast to American PDAs, which mostly featured touchscreens, Psion organizers had generously sized, built-in keyboards, and by 1987 they had their own operating system, called EPOC. Early Psion models resembled calculators; later ones resembled tiny laptops.

Psion organizers quickly attracted a cult following for their portability, wealth of third-party apps, and computing capabilities, which included word processing and spreadsheet programs. "They were really serious computers and a favorite of people in the early Web community, such as Tim Berners-Lee,"* says Marc Weber, a curator at the Computer History Museum, who has studied the history of mobile devices. Cellphone companies took note of Psion's success. Ericsson and Nokia knew they needed high-quality mobile software for their smartphones. They also wanted phone makers to control their own operating systems rather than outsource them to outsiders like Microsoft. When Psion floated the idea of a partnership, they were receptive. Rydbeck says: "We realized that soon software would be a whole area of its own, not just a little thing on the side. Microsoft was beginning [mobile software work] and suggested we could partner with them. But we thought cellular companies might be able to own this technology."

In June 1998, Psion spun off its software division to create a joint venture, called Symbian. Psion became the largest shareholder and viewed the deal as a way to reduce its operating costs and gain access to the emerging smartphone market. Ericsson and Nokia, which originally each took a 30 percent ownership, got a tried-and-tested,

---

*The British computer scientist who is credited with inventing the World Wide Web.

mobile-specific operating system independent of Microsoft's control and high licensing rates. Motorola formally joined Symbian in October 1998, and Japan's Matsushita (Panasonic) followed in 1999.

Symbian became a London-based company that produced an operating system and basic apps that its owners and other companies could license. It was Symbian's job to develop software components that enabled a range of smartphone services—everything from a phone's communications to its multimedia support to its core, pre-loaded applications. Symbian then packaged these components into a few "reference designs" for phone makers, who selected the reference design they liked, licensed the software, integrated it into their hardware, and paid Symbian $5 for every smartphone they shipped. Symbian also created SDKs, so third-party developers could build apps for Symbian phones.

The Symbian founders considered themselves a team of equals, but Nokia "quickly moved ahead and became the 'big brother' in the team," says Rydbeck. The power shift reflected Nokia's strength in the market, where it was taking share from both Motorola and especially Ericsson. Compared to its longtime European rival, Nokia had more phone models, a better-known brand, and lower costs, due to its larger size. It was also the first company to launch a complete version of Symbian that was fully compatible with third-party applications, content, and services. This new version appeared in 2001 on its third-generation 9210 Communicator.

As Nokia widened its lead as the world's largest cellphone and smartphone maker, its competitors gradually withdrew from Symbian, an irony, since Symbian itself had been the cellphone industry's response to another company's attempt to dominate the market—Microsoft. Microsoft had entered the mobile software market in 1996, when it started selling manufacturers a lightweight version of its Windows operating system called Windows CE. Microsoft had made several earlier attempts to develop a mobile version of Windows but, dissatisfied with the results, had scrapped the projects. But in what *Businessweek* observed as "a classic example of how Microsoft stubbornly pushes its way into new fields, even if it takes years,"[38] Microsoft continued to refine what became Windows CE and convinced a number of big companies, including the American PC makers Compaq and Hewlett-Packard (HP), the Japanese electronics maker Casio, and the European electronics maker Philips, to adopt it and make handheld PCs, which resembled small laptops. The software enabled portable use of condensed versions of Microsoft Office apps so business travelers could work while away from their PCs.

Microsoft spent the next decade adding features to Windows CE and making it compatible with more devices. In 1998, Microsoft released a version that device makers could use for what it called Palm PCs—smaller, lighter, touchscreen PCs that users could operate with one hand. Though they were bulkier and more expensive, they were Microsoft's answer to Palm's Pilot PDAs, right down to their name. Inevitably, Palm sued over the name, forcing Microsoft replaced that operating system with software called Palm-size PCs. In 2000, Microsoft redesigned Windows Pocket PC. Pocket PCs retained the organizer features of the earlier handheld and Palm PCs but had a simpler user interface. They also added support for multimedia, including digital MP3 songs, video clips, electronic books, and audio books.

Though a few companies produced Windows-based PDA-phones as early as 2001, the first device to be widely recognized as a "Microsoft-powered smartphone"[39] was the Orange SPV. A collaboration between Microsoft, Taiwan's HTC, and the European carrier Orange, the SPV was launched in Europe in 2002 and used Microsoft's Smartphone 2002 software, which was designed to challenge Symbian and the Palm OS and included Microsoft Outlook's e-mail and calendar features as well as slimmed-down versions of Internet Explorer, Microsoft's MSN Messenger instant messaging service, and Windows Media Player. The *Wall Street Journal* criticized the small (4.3 inches by 1.78 inches by 0.78 inches; 3.35 ounces) candy bar–shaped phone for being "packed with a bewildering array of features, many of them cool, others clumsy,"[40] and users complained about the phone's short battery life and the software's tendency to freeze, but a low price ($278) and a $20 million joint marketing campaign helped sell 50,000 of the phones in Britain within six months.[41] A year later Microsoft adopted the name Windows Mobile for its mobile software, noting that "Windows is a brand customers associate with powerful and familiar software,"[42] and continued updating it with new features and improvements on a close to annual basis.

I-MODE AND JAPAN

In the early 2000s, Microsoft owned Windows Mobile, Nokia (mostly) owned Symbian, the reintegrated Palm (which acquired Handspring in 2003) owned Palm OS, and RIM owned its suite of BlackBerry technology. No one owned the mobile Web—except in one country. In Japan, a single company dominated the mobile Web for years. That company was a carrier: NTT DOCOMO.

Though it wasn't always obvious to the rest of the world, Japan has been a wireless technology leader for decades. Japan was the first country to launch a 1G network, in 1979, and the first country to launch a commercial 3G network, in 2001. Japan's Sharp was the first phone maker to integrate a camera into a cellphone, in 2000, and Japan was one of the first countries to embrace the use of cellphones as mobile wallets, in 2004.

Japan also sparked the world's first mobile media revolution, in 1999, via DOCOMO's mobile Web service, i-mode. DOCOMO was early to understand that people wanted quick, easy access to lots of online content on their phones. Browsing the Web via phone has never been easy. "The way the Web developed was not friendly to lower-powered, mobile devices," says Marc Weber. "The Web was primarily designed for big computer monitors, so early smartphones needed to resize Web pages and make graphics simpler." DOCOMO didn't solve all of these problems, but with the help of a Japanese software firm called ACCESS, it figured out how to maximize the technology available at the time.

In its early years, i-mode ran on small flip phones and featured a dedicated "i" button. Pressing the button launched a browser that loaded DOCOMO's i-mode portal. The portal linked to hundreds of websites written in compact HTML, a simplified version of the code that developers use to add basic design elements to Web pages and to link pages to each other. Compact HTML streamlines websites for better viewing on cellphones that have small screens and limited memory while still allowing sites to display simple graphics. websites had to be rewritten in compact HTML to display properly in the i-mode browser, but compact HTML's similarity to HTML made the process relatively simple and affordable. This was one of i-mode's strengths when compared to WAP, which Europe and the United States used to translate the Web to phones; WAP sites used a different, less familiar programming language, and thus took longer to create. ACCESS developed compact HTML and the i-mode browser and supplied both to DOCOMO.

Within six months of launch DOCOMO had 116 "official" i-mode sites that it listed on its i-mode portal and more than 1,100 unauthorized sites that users could visit by typing in the Web address.[43] By November 2000, i-mode boasted 23,000 official and unofficial sites, while WAP sites numbered less than 1,000.[44] The i-mode universe felt nearly as big as the regular Internet—and users didn't have to dial up to see these sites. Unlike WAP, the i-mode service was always on.

DOCOMO didn't just convince people to go online regularly from their phones. It also got people to gladly pay to be online. Low rates and convenient billing were the key. Users paid small fees for access to i-mode, for the amount of data they consumed while online, and for any content they purchased from i-mode sites. In 1999, i-mode–equipped phones cost between $100 and $350 each, the service's monthly access fee was $3, and a user's average monthly spending was about $19.[45] DOCOMO added the fees to subscribers' monthly phone bills and acted as a money middleman. It paid i-mode content providers on behalf of the users and charged the companies 9 percent for processing. Since using i-mode didn't feel onerous or expensive, people quickly became active, paying customers. As a result, i-mode–hosted services, ranging from daily horoscopes to online banking, flourished and multiplied. Like the sale of smartphone apps today, the system was a positive feedback loop in which success fed success.

By 2001, DOCOMO was Japan's largest company by market capitalization.[46] By early 2002, one in four Japanese people used i-mode.[47] Outsiders often attribute its traction to cultural conditions, citing a Japanese predilection for technology and shopping. Tomihisa Kamada, who led i-mode technology development at ACCESS, says it was actually Japan's advanced wireless networks and carrier-centric phone industry that enabled it to take off. I-mode technology was fast enough to offer instant gratification, and DOCOMO wielded enough power over phone makers and content providers to establish a cohesive phone/mobile content/billing ecosystem.

Japan was perhaps the only country where carriers had such wide-ranging clout; when DOCOMO tried to expand i-mode to Europe, the United States, and some Asian countries, the system flopped. I-mode is also a prominent example of the ways Japan's cellphone industry tends to be innovative yet insular. The Nomura Research Institute, Japan's largest information technology consultancy, named this the "Galápagos effect." Like the exotic animals and plants that evolved on the Galápagos Islands, phones and mobile services developed for the Japanese market are often unique in the world—highly adapted to their own environment but incompatible with other regions.

Though i-mode did not flourish globally, its domestic success taught smartphone companies around the world significant lessons. It showed that consumers wanted to use the mobile Web and would pay for access and content, provided they saw value in it. I-mode phones demonstrated how to create a Web-centric user experience even without a modern smartphone operating system. Ryuji Natsuumi, an ACCESS executive who developed the i-mode browser along with Kamada, says, "Without the i-mode handset, the evolution to the smartphone would have taken longer."

I-mode also illustrated the need to offer consumers a wide variety of mobile Web content, and proved that a single company could foster a mobile content ecosystem capable of growing at a viral rate. These ideas cropped up again in the iPhone. The iPhone would deliver a real Internet experience rather than a trimmed-down, mobile-only one, and formed its own robust ecosystem, with Apple firmly in charge.

By the end of 2002 the first generation of smartphones had launched and found audiences. BlackBerrys claimed much of the business market and other users who needed fast, secure e-mail access. Microsoft was getting up to speed with its smartphone software, which attracted both business users and consumers. Treos appealed to people who were used to Palm devices and wanted user-friendly software. Symbian phones held sway in Europe through Nokia and Sony Ericsson, a London-based joint venture; Sony and Ericsson formed in 2001 by merging their cellphone businesses. I-mode phones had captivated Japan with their easy mobile Web access and moved millions of people online.

Most of these first-generation smartphones had real operating systems. They would soon be given color screens, cameras, Bluetooth[46] and Wi-Fi[47] and Global Positioning System (GPS)[48] connections, and music and video capabilities, if they didn't already have them. Apps were still relatively simple, though, touchscreens couldn't recognize more than one finger at a time, and the mobile Web was usually a lesser version of the regular Web. This would remain the status quo until the iPhone upended everything.

2

Apple, Google, Microsoft, and Samsung

APPLE

"We're going to make some history together today." With those now famous words, Apple CEO Steve Jobs took the stage to debut the iPhone at the annual Macworld conference in San Francisco on January 9, 2007. After running through some other announcements, Jobs pulled an iPhone out of his jeans pocket. The phone's design was striking. It had a glossy, 3.5-inch high-resolution screen dotted with colorful icons and surrounded by a shiny, stainless-steel rim. At 4.5 inches by 2.4 inches by 0.46 inches, the iPhone was also thinner than most other phones on the market.

Original iPhone (far left) and newer iPhones (*Yutaka Tsutano/Flickr*)

Jobs spent the next 90 minutes electrifying his audience as he outlined the iPhone's features. There was the "revolutionary [user interface], the result of years of development"; a "supersmart," multitouch screen that could discern gestures and "works like magic"; an operating system "five years ahead of what's on any other phone"; iPod software that let users "touch your music"; a virtual keyboard that's "really fast to type on"; and the "first fully usable browser on a cellphone." [1] Apple, said Jobs, wanted to "leapfrog" current smartphones and create "a product that's way smarter and easier to use." To punctuate his point Jobs displayed a photo of some leading smartphones. The picture included a Palm Treo, a BlackBerry Pearl, a Symbian-powered Nokia E62, and a Windows Mobile–powered Motorola Q. "After today, I don't think anyone's going to look at these phones in the same way again," Jobs said, and although his presentation was laden with hyperbole and showmanship, he was right. Next to the iPhone, these high-end phones, which had comparatively small screens and built-in QWERTY keyboards, suddenly seemed passé.

Jobs hadn't always been so enthusiastic about smartphones or supermobile computing. He didn't like the way the carriers controlled the smartphone business, and he was acutely aware that the smartphonelike Newton had been a failure. Jeff Hawkins recalls approaching Apple in 2003 with Ed Colligan and Donna Dubinsky to discuss a possible collaboration with Handspring. Jobs wasn't interested. As Hawkins tells it: "Steve goes up to the whiteboard and starts drawing. He says, 'At the center of the world is your Mac, and then there's music and video and this stuff. And there's a teeny little circle over here, which is a handheld. Maybe some people have that. It's not really very important.'" Hawkins had a different opinion. He leaped up to the board, drew a handheld computer, and told Jobs, "No, Steve, I disagree. *This* is going to be the center of your world." Hawkins, Colligan, and Dubinsky think the meeting influenced Jobs to get into the smartphone business.

In his biography of Jobs, Walter Isaacson writes that the catalyst was cellphones' encroachment on iPods, which was obvious by the mid-2000s. According to Isaacson, Jobs told the Apple board of directors, "Everyone carries a phone, so that could render the iPod unnecessary." [2] In 2005, Sony Ericsson introduced a line of Walkman music phones, and Nokia announced its intention to sell music phones branded XpressMusic, showing Jobs he was right to be wary of phone makers. Phones in both series had built-in MP3 music players.

Besides the music-phone threat, Jobs came to see cellphones as a large and lucrative new market. Apple was selling tens of millions of iPods a year, but cellphone makers were selling hundreds of millions of phones.[3] Jobs started talking to Motorola about putting an iPod inside one of its cellphones. In 2005, the two companies unveiled the music-focused ROKR E1—the first cellphone that let people download music from iTunes, transfer it (via a computer and USB cable) and listen to it on an integrated iTunes music player. Dubbed the "iTunes phone," it generated a huge amount of attention but was poorly received, due to its clumsy (non-Apple) software, sluggish response time, limited storage, and boring design. Jobs knew Apple could do a better job. As Phil Schiller, Apple's marketing chief, explained in 2012: "At that time, cellphones weren't any good as entertainment devices. That made us realize—maybe we should make our own phone."[4]

Apple's efforts to develop what would become the iPhone were shrouded in secrecy. Abigail Sarah Brody, a former Apple designer who contributed to the phone's software, says the work was mysterious, exciting, and demanding. Brody had been at Apple since 2001, mostly designing the professional photo, video, and music editing applications Apple sells to Mac computer users. One day, in 2005, Brody was invited to become part of a team of three that included an engineer and an executive—one of a number of small groups Apple set up to assist with the iPhone's design. She accepted. The trio's assignment was to create a user interface for an upcoming device. All they were told was that the device used a "multitouch" screen.

Before the iPhone, few people were familiar with multitouch. Regular touchscreens, such as the type used in early smartphones and ATMs, can't recognize gestures, because they can't accurately decipher contact points from more than one finger or hand at a time. Research on multitouch technology began in the 1980s, but the specific kind used in the iPhone was mostly created in the late 1990s and early 2000s. That's when a Delaware start-up called FingerWorks began leveraging research from the University of Delaware's computer and electrical engineering department. According to a 2004 press release,

FingerWorks's technology consisted of "hardware and software elements for sensing, tracking, and interpreting the motion of multiple hands and multiple fingers on a touch imaging surface."[5] Apple, which had also developed some multitouch technology in-house, acquired FingerWorks and its technology in 2005.

After Apple "pulled" Brody into "the secret iPhone project," as she describes it, she was given a multitouch screen so she could see what it entailed. It was "a crude prototype," but she guessed she was working on a future phone. "After seeing the iPod, it was very logical that the next thing for Apple to make was an iPhone," she explains.

To determine what size the phone's icons should be, Brody measured how far her thumb could reach across the screen when she held the prototype. She asked the engineer in her team to do the same, since he had larger hands. "It was very simple; just using our bodies as a means to measure," she recalls. "We wanted to have proper ergonomics and consistency in the user interface." For the next six months Brody's team focused on creating supremely user-friendly software.

Apple won kudos for the effort it put into the iPhone's design, and it was an instant hit. People had speculated for years about a possible iPhone, but most thought it would be a cellphone infused with iPod features, like the ROKR E1. Apple easily exceeded expectations. Some reporters expressed skepticism about Apple's foray into a highly competitive industry and the iPhone's comparatively high price ($499 to $599, depending on the amount of storage capacity). But most raved about the device and Jobs's thrilling presentation. As the *New York Post* put it: "Apple Cells It—Breakthrough iPhone Wows High-Tech Fans and Wall St."[6]

Post-Macworld, Apple shares hit a then record high, while those of competing phone makers, including Motorola, Nokia, Palm, and RIM, slumped. Online searches for "iPhone" spiked as consumers clamored to learn more about the device. A Deutsche Bank analyst described initial customer interest as nothing less than "rabid."[7]

The iPhone went on sale on June 29, 2007. Over the next year Apple sold 6.1 million. A $200 price drop (from $599 to $399) that Apple instituted in early September spurred purchases in the last quarter of

2007, though the move initially angered the early adopters who had bought the phone at full price. (Apple appeased those customers by giving them $100 in Apple store credit.) In 2008, Apple sold 14 million iPhones, handily beating its sales goal of 10 million for that year.

Not everything about the iPhone was as revolutionary as Jobs claimed. The camera was just two megapixels, had no flash, and couldn't record video. The phone had no GPS, didn't synch with Microsoft Outlook's corporate e-mail client, and couldn't edit Microsoft Office documents. Most controversially, the iPhone was only available on AT&T and ran on its EDGE network instead of 3G. The disparity between EDGE, dubbed 2.75G, and 3G was mostly one of speed. EDGE was about three times faster than GPRS but two times slower than 3G technology, which was live in 160 cities across the United States by the time the iPhone went on sale. The allure of 3G was mobile multimedia—the ability to quickly load Web pages, download music, and even stream online videos to their phones. A 2007 *New York Times* article noted, "Early reviews of the iPhone, while positive, have faulted the slower [EDGE] network because it will limit the palm-size wireless computer's greatest strength—making the Internet easily accessible on the go." [8]

Other phones at the time did support 3G, but where the iPhone leapfrogged over other smartphones was its user experience. It was a phone designed to serve users first, rather than carriers. People loved navigating their iPhones by tapping, swiping, and pinching their fingers. The *Guardian* called multitouch "the soul of the iPhone," [9] because it affected everything the user did on the device. Compared to other smartphones, the iPhone's software felt speedy. "In typical smartphones at that time, the user interface was sluggish," says former Nokia manager Jari Kiuru. "After pressing a button you'd have to wait for things to happen. The iPhone had a fast reaction." The iPhone even seemed intuitive. Embedded sensors could detect ambient light conditions and the phone's physical orientation and proximity to a user's face. The sensors dimmed the phone's screen accordingly and shifted its icons when the user rotated the phone. The device almost felt alive to former Ericsson CTO Nils Rydbeck. "Apple made the iPhone human, like an extension of your senses," he says. "Steve Jobs moved smartphones from a technical mess to a thing you love."

The iPhone also figured out how to translate the desktop Web experience to a much smaller screen. In his Macworld presentation Jobs characterized the iPhone as a "breakthrough Internet communicator" that "put the Internet in your pocket for the first time ever." This was another instance in which he wasn't exaggerating. On the iPhone the mobile Web seemed less cumbersome and limited, even without 3G. Instead of a simplified WAP mobile browser, the iPhone ran a mobile-optimized version of Safari, the Web browser Apple developed for its Mac computers. Through Safari, iPhone users could access the entire Web as it was originally designed, not just sites adapted for smartphone viewing. There was no laborious scrolling. Instead, users touched the screen to zoom in and out of sections they wanted to read. Safari also let users open multiple websites and switch between them, like on a desktop computer. Historian Marc Weber calls the iPhone "the first usable device for the mobile Web. . . . The iPhone wasn't the first smartphone, but it was the first phone really designed as a primary way to access the Web regularly."

The other thing the iPhone got right was timing. By the time the iPhone debuted, mobile Internet access was critical for many people. The iPhone met this need and went on sale just as networks were getting fast enough to support Web-centric smartphones. The first iPhone may not have fully leveraged top network speeds, but subsequent iPhones did. Wi-Fi was pervasive by 2007, giving users another connectivity option when AT&T's network slowed. People had their home and office Wi-Fi and the 11,000 public hot spots AT&T operated nationwide, including in airports, Barnes & Noble bookstores, and McDonald's restaurants.[10] Apple also capitalized on the fact that consumers had started buying smartphones for personal use. The iPhone's potential audience was much broader than the business elite.

Apple's timing caught phone makers flat-footed. At the time, RIM, Palm, Motorola, and Nokia led the U.S. smartphone market, in that order. None of them had anything equivalent to the iPhone. Most of them downplayed the threat in order to reassure their investors, and also because it was not yet apparent the iPhone would succeed. Padmasree Warrior, who was then Motorola's CTO, published a blog post on Motorola's website the day after Jobs's presentation. After complimenting the iPhone's design, Warrior noted, "There is nothing revolutionary or disruptive about any of the [iPhone's] technologies."[11] Eight days after Macworld, Microsoft CEO Steve Ballmer scoffed at the iPhone's design and price during a CNBC TV interview. "That's the most expensive phone in the world," he said, "and it doesn't appeal to business customers, because it doesn't have a keyboard."[12] Around the same time, the *Korea Times*, an English-language South Korean newspaper, quoted an anonymous Samsung official as saying, "Although we are waiting to see how U.S. consumers will react, we are not impressed by [the iPhone's] features."[13]

In a conference call with analysts later that month, Nokia CEO Olli-Pekka Kallasvuo said the iPhone would "be good for the industry" and wasn't "something that would in any way necessitate us changing our thinking [about] our software and business approach."[14] RIM's co-CEO Jim Balsillie made similar remarks in June 2007. "[Apple] did us a great favor, because they drove attention to the [smartphone] space," said Balsillie. "We think that the attention to [the iPhone] . . . has, quite frankly, been overwhelmingly positive to our business."[15] Ed Colligan, who by that time was Palm's CEO, had a more measured response. In an April 2007 analyst call, Colligan said, "I think [the iPhone] will reach a slightly different [customer] segment [than] what we're going after." But he added, "We're taking it as a serious competitor [and] competitive threat."[16]

Yet to smartphone pioneers, including Hawkins, Canova, and Rydbeck, the iPhone felt like validation. Hawkins had already departed Palm to pursue his true passion, neuroscience, and subsequently founded a data analytics start-up called Numenta. He saw in the iPhone the kind of phone he tried to make in the early 2000s. "Apple broke the floodgates open," Hawkins says. "Now we had a true handheld computer that wasn't constrained by carriers." When Canova first saw the iPhone he remembered the Neon prototype he built for IBM and that phone's tilt sensor-equipped screen. Canova says he thought, "Wow, they did it. I had this idea way back, but they managed to do it." Rydbeck, too, felt a deep sense of satisfaction: "I thought, 'Finally, this product is right; finally, a guy who understood exactly what we wanted to do.' I was as happy as if it were my own company."

Nevertheless, the iPhone wouldn't completely fulfill their vision until it supported third-party applications, a fundamental smartphone qualification. As some research firms noted at the time, the first iPhone's lack of this capability technically classified it as a very high-end feature phone. "While the iPhone is undoubtedly clever and capable, it is not correct to call it a smartphone," concluded a January 2007 ABI Research commentary.[17]

Other smartphones had incorporated apps from outside developers for years, but Jobs opposed the idea. As Isaacson's biography recounts: "[Jobs] didn't want outsiders to create applications for the iPhone that could mess it up, infect it with viruses, or pollute its integrity."[18] When he first unveiled the iPhone, Jobs clearly viewed it as a closed device. In an interview with the *New York Times*, he said: "[Apple] define[s] everything that is on the phone. You don't want your phone to be like a PC. The last thing you want is to have loaded three apps on your phone, and then you go to make a call and it doesn't work anymore. [iPhones] are more like iPods than they are like computers."[19]

According to Isaacson, Apple executives and board members were early to recognize the value of third-party iPhone apps. In the hopes of swaying Jobs, they enumerated the benefits that consumers and Apple would reap from having a wide array of them. Jobs's inner circle also stressed the competitive angle. Other smartphone companies, such as Microsoft and Nokia, were courting developers for their platforms. Jobs's advisers warned him that if Apple didn't start recruiting too, developers would end up supporting Apple's rivals.[20]

By mid-2007, Jobs had softened his stance. In June Apple started encouraging developers to write apps for the iPhone using Web 2.0 standards. This meant developers could employ recent versions of Web coding languages, such as HTML, JavaScript, and CSS, to create Web apps that would run inside the iPhone's browser, rather than native iPhone apps that users installed on their home screens and could remain active in the phone's background while users did other tasks.

Apple's policy eased the way for a broad community of Web developers to create iPhone apps, since they wouldn't have to learn the intricacies of the phone's operating system. But the policy also meant iPhone apps would have limited functionality because they would only be active inside browsers and would not work when users were offline or had poor network connections. Developers were vocal about their displeasure. *InformationWeek* observed, "Even though Jobs says Web-based apps 'look and behave exactly' like native apps, some Mac software developers don't buy it. Words like 'lame' and 'weeeeak' [have] popped up in response on blogs and message boards."[21]

Jobs eventually embraced a compromise: the iPhone would support native apps created by outside developers, but Apple would vet all apps and distribute them exclusively through iTunes to ensure quality. In October 2007, Apple announced it would release a software development kit in coming months. Apple also posted a message on its website, signed by Jobs, that said, "We are excited about creating a vibrant third-party developer community around the iPhone and enabling hundreds of new applications for our users."[22] Apple delivered a beta version of its promised SDK in March 2008 and a final version in July.

The Apple App Store opened on July 10, 2008, 13 months after the first iPhone went on sale and in time for the new iPhone 3G, which Apple billed as "twice as fast at half the price,"[23] because it ran on 3G networks and cost $199 to $299. When Jobs introduced the App Store, he told *USA Today*, "We think there's never been anything like it."[24] This was, again, a slight exaggeration. Independent app stores such as Handango had been around since the late 1990s or early 2000s, initially distributing PDA apps and later selling ones for BlackBerry, Palm, Symbian, and Windows Mobile. In 2008, Handango offered more than 16,000 smartphone apps through its app portal.[25]

Apple centralized its app operations tightly around iTunes and the iPhone. It not only built the App Store into iTunes, but also built the store directly into the iPhone, enabling users to quickly discover and enjoy iPhone apps. People simply purchased them with a click—or selected them in the case of free apps—and downloaded them wirelessly over cellular networks or Wi-Fi. At the time, Handango and a company called Handmark had partnerships with some carriers and smartphone makers that placed their app software on select Black-Berry, Palm Treo, Symbian, and Windows Mobile phones. These on-device storefronts let consumers browse and download apps over the air like the App Store, but the latter soon passed them in quality and diversity. Apple's strict standards and comprehensive software tools, combined with the iPhone's large, bright screen and intuitive user interface, produced apps that looked better and did more than those of other smartphones.

When the App Store debuted, Jobs called it the "biggest launch of my career."[26] The store boasted more than 550 apps, including big names like eBay, Electronic Arts, Facebook, Major League Baseball, the *New York Times*, Oracle, and Salesforce. In the first 72 hours after its launch, consumers downloaded 10 million apps. Jobs called the results "staggering" and "a grand slam."[27]

By December 2008, the App Store stocked more than 10,000 apps. Many of them were so entertaining that *USA Today*'s Ed Baig compared browsing apps in iTunes to "trolling through the aisles of a toy store."[28] Games proved to be the most popular, as they still are.

Smartphones have hosted games ever since the IBM Simon. After RIM gave BlackBerrys color displays in 2003, it started loading a game called BrickBreaker onto them. BrickBreaker was just a simple clone of Atari's Breakout arcade game and involved little more than bouncing a ball into a layer of bricks to destroy them. Nevertheless, Brick-Breaker amassed a loyal following of tense executives who depended on it for stress relief. Between 2003 and 2010, Nokia tried to bring a video game console–like experience to smartphones with a mobile gaming platform called N-Gage. N-Gage enabled users to download popular titles such as soccer games branded by FIFA, the international governing body of the sport, onto their phones and compete against other people online. But since N-Gage never achieved mass popularity, the iPhone is widely considered the first smartphone that doubled as a gaming machine.

At the end of the App Store's first six months, Super Monkey Ball from the Japanese video game publisher Sega was one of the bestsellers among paid apps. The game cost a steep $9.99 and involved guiding a bubble-encased cartoon monkey through mazes. Another game, Tap

Tap Revenge, which measured how accurately users could tap out a song's tempo on their phone screens, attracted even more downloads, because it was free. Social networking, Internet radio, and location-based apps were also highly popular.

The iPhone had become a true handheld computer with a broad range of potential features, just like IBM, Palm, and others had imagined years earlier. With the introduction of the App Store, Apple formed its own, self-sustaining ecosystem between consumers, developers, and the iPhone. Like Japan's i-mode, the iPhone was really a software and services framework tied to a phone, and it was the iPhone ecosystem that pushed the entire smartphone industry forward. Once again, Apple had caught other smartphone companies on the hop. Within a few months, however, a new rival would emerge, presenting Apple with its first real competition.

ANDROID

Unbeknownst to Apple, Google began developing its own smartphone in 2005. The project stemmed from a 2002 meeting at Stanford University between Google's cofounders, Larry Page and Sergey Brin, and Andy Rubin, an engineer with a deep interest in robotics and mobile technology.[29] Rubin began his mobile software career in the early 1990s, when he helped develop the PDA operating system and user interface Magic Cap at General Magic, a start-up Apple had spun out as a separate business. Several big-name companies, including Motorola and Sony, produced Magic Cap PDAs, which won attention for their sophisticated graphics and built-in wireless data connections but struggled in the marketplace.

Rubin drew upon that experience in 1999 when he cofounded a start-up, Danger Research Inc., which designed a smartphone platform called hiptop and a trendy series of smartphones that T-Mobile sold under the name T-Mobile Sidekick. Rubin wanted Sidekicks, or hiptops, as they were known internally, to be consumer versions of BlackBerrys—cheaper and more entertaining but similarly data-centric.[30] When T-Mobile began selling Sidekicks in October 2002, they were one of the first smartphones to feature always-on cellular data—the GSM/GPRS connections RIM and Handspring had raced to incorporate into their devices. Danger leveraged two-way data to let users send and receive e-mail and chat over instant messaging services anytime, anywhere.

Original Danger Hiptop/T-Mobile Sidekick (*David Mueller/ Wikimedia*)

Sidekicks spearheaded a number of other significant smartphone features, including an on-device app store. Starting in 2004, Sidekicks shipped with a built-in program called the Download Fun Catalog that featured apps, wallpapers, and ringtones, complete with descriptions and screenshots, and enabled users to download them straight to their phones. The charges showed up on their phone bills. "We knew that third-party software would be critical to our success, so we made it easy [to buy and sell apps and install them]," explains Chris DeSalvo, who was a senior software engineer at Danger from 2000 to 2005. "No extra setup, no hassle." Though more limited in scope than the iPhone's App Store, the Download Fun Catalog had the same basic functionality, four years earlier, and people actually used it. DeSalvo says some Sidekick developers made $40,000 or more per month off app sales in the catalog's early years.

Sidekicks were also one of the first phones that let users wirelessly synchronize and save their phone data to an online account. Danger's cloud storage system automatically saved everything users did on their hiptops, which enabled them to manage far more data, such as e-mails, than their phones would have been able to hold on their own since smartphones didn't have much storage capacity at the time. "By having our servers hold it all we enabled the hiptop to dynamically flush old stuff out and pull new stuff in, knowing that we could always do another shuffle later on, if you needed to see that old stuff again," says DeSalvo. This service had hiccups, including a 2005 hack of personal data from socialite and Sidekick user Paris Hilton's account and a 2009 server outage that erased user data. But many consumers liked the idea of backing up their phone data in case of phone loss, theft, or damage and other smartphone companies later introduced similar services.

Though Sidekicks blazed a trail, they were never broadly popular and still don't get much recognition for their innovations. DeSalvo thinks distribution, marketing, and most of all timing curbed the phones' sales—and, consequently, their legacy. "We were just too early," he says. "We had all this great stuff, and no one had any idea that they wanted or needed it." It didn't help that T-Mobile largely marketed Sidekicks to teens. "The hiptop was presented as a cutesy fashion accessory for kids rather than a powerful tool that could unchain you from your desktop computer," laments DeSalvo.

In 2003, Rubin left Danger and cofounded Android Inc. The initial plan was to create a software platform that would let users move photos off their digital cameras and store them in the cloud, but the start-up decided to develop an "open-source handset solution" after realizing phones were a far larger target market than cameras.[31] Larry Page, who was interested in mobile technology, had kept in touch with Rubin following their meeting at Stanford, and in 2005, the two met to discuss Android. Soon after, Google quietly acquired the start-up, reportedly for $50 million,[32] making Android a wholly owned subsidiary.

Still, Android operated so stealthily that rumors of a Gphone didn't surface for more than a year. In December 2006, *The Observer*, a British newspaper, reported that Google had met with Orange to discuss a "branded Google phone" with "built-in Google software."[33] In March 2007, a Boston-based technologist with "an inside source" blogged that Rubin was building a "BlackBerry-like, slick device" with a 100-person team.[34] Speculation escalated, with many believing Google was creating a no-cost, completely ad-supported phone. In September 2007, the *Los Angeles Times* noted, "The Google phone is like the Roswell UFO: Few outsiders know if it really exists, but it has a cult following."[35]

Google tried to quiet this speculation in November 2007 by announcing the creation of Android and the Open Handset Alliance. Google characterized Android as "the first truly open and comprehensive platform for mobile devices,"[36] because it would be free and available to anyone who wanted to use it, and could be customized. The alliance was a coalition of 34 technology companies, including Google, other software providers, manufacturers, and carriers that would collaborate on Android with a "common goal of fostering innovation on mobile devices."[37] Google executives emphasized that the new operating system, based on the free, open-source software Linux,* would "power thousands of different phone models."[38] This disappointed those anticipating an iPhone-like Gphone. An *Advertising Age* article bemoaned "the fact that the Google Phone turned out to be not an actual phone but just a giant nerd committee."[39]

The public wouldn't see the first Android phone until September 23, 2008, when a launch event was held in New York City. By that time people knew the mythical Gphone was real, because most of the major details had leaked, including the phone's name (G1), manufacturer (HTC), exclusive U.S. carrier (T-Mobile), and basic design

---

*Linux software was initially developed as a PC operating system and now runs everything from computer servers to supercomputers to gadgets, and is supported by the donated efforts of thousands of programmers around the world.

(large touchscreen with slide-out QWERTY keyboard and a trackball for navigation). As expected, the G1 integrated multiple Google properties, such as Gmail, Google Calendar, Google Maps, Google Talk instant messaging, Google Search, and YouTube. A speedy, full-fledged Web browser and the user's Google account tied everything together. "Suddenly, the hodgepodge plethora of Google applications and services have a single point of connection and synchronization," reported *eWEEK*.[40]

T-Mobile G1, the first Android smartphone (*Elizabeth Woyke*)

The G1 went on sale in late October, priced at $179, $20 less than the iPhone. Early reviews tended to applaud Android, find fault with the phone's hardware, and critique T-Mobile, which was still rolling out its 3G network nationwide. David Pogue, the former *New York Times* technology columnist, gave the G1's software an A– ("smartly designed"), the hardware a B– ("thicker, heavier and homelier than the iPhone"), and T-Mobile a C ("one of the [country's] weakest networks").[41]

Comparisons to the iPhone were inevitable. A *New York Daily News* headline read, PHONE FIGHT! T-MOBILE TEAMS WITH GOOGLE TO TAKE ON APPLE'S KINGPIN.[42] But the iPhone easily beat the G1 on most counts. Virtually no one preferred the G1's bulky design to the iPhone's unless they were big fans of built-in keyboards. Google's mobile app store, then called the Android Market, also rated low marks. It debuted on the same day as the G1 with far less polish than the App Store. The *Guardian* described the market's assortment of about 50 apps, few of which came from big brands, as "desperately thin."[43]

Unbeknownst to the public at the time, the G1 was Google's Plan B. Google originally aimed to introduce its first Android phone in late 2007 but postponed the release after seeing Jobs's unveiling of the iPhone. As journalist Fred Vogelstein details in his book *Dogfight: How Apple and Google Went to War and Started a Revolution*, Rubin and his team intended to launch Android with a small-screen, QWERTY-keyboard phone, but when they saw the iPhone's multitouch screen, they realized they too needed to offer a touchscreen and shifted their efforts to a different phone that became the G1.[44]

Google's governing idea for Android remained the same: to create an operating system that would appeal to anyone who used Google services. DeSalvo, who joined Android as a senior software engineer in 2005, says: "We wanted the G1 to be a seamless extension of the relationship you already had with Google [on your computer]. You should be able to start a draft Gmail on your desktop, walk away, pick it up later on your G1, finish editing, and send it. A chat you were having on your desktop in Google Talk should be right there on your phone, [too]."

The iPhone couldn't do this type of fluid, over-the-air data syncing, but the iPhone's stylishness made the Android team nervous. "The scrolling [on the first iPhone] was supersmooth, the colors were great, the icons and typography were gorgeous," DeSalvo recalls. "Visually, the software we had at the time was nowhere near as good." The contrast between the two phones and platforms was obvious: the iPhone was "an extremely beautiful realization of a very limited device" while

Google's Android prototype was "a much uglier realization of a very powerful device," he says.

The Android team knew first impressions mattered to consumers. To compete, Android would have to exploit features the iPhone couldn't or wouldn't offer: tight coupling with a user's Google accounts and broad uptake across multiple carriers. "Google already had insane numbers of users. If we could convince just a few percent of them to get an Android phone, then we'd be a huge success," adds DeSalvo.

Android took time to catch on, and the G1 wasn't an iPhone-level hit. The first iPhone sold 1 million units in 74 days. The iPhone 3G passed that mark in three days. It took the G1 about six months to cross that line. But its integration of Google services, its speed, and the ability to customize it gave the G1 a unique appeal. CNET wrote, "In a battle of pure looks, Apple's iPhone would win hands down, [but] the real beauty of the T-Mobile G1 is the Google Android platform, as it has the potential to make smart phones more personal and powerful." [45]

Like the iPhone, Android borrowed from its predecessors. The G1's large touchscreen was an obvious reference to the iPhone, but the phone also had hiptop DNA. Its slide-out keyboard resembled the Sidekick's, and not by coincidence. "All of us loved our hiptop keyboard," DeSalvo says. "We had put a lot of work into its design [and figured out] what makes a really great thumb keyboard."

Android also incorporated software notifications, which had been another hiptop innovation. Prior to Sidekicks, most mobile devices besides BlackBerrys didn't have a good way to tell users they had new voicemail or remind them of an upcoming appointment. On Sidekicks, notifications subtly appeared at the top of phone's screen to inform users of missed activity. For the G1, DeSalvo and his colleagues improved on that system by adding a virtual "window shade" that listed notifications, which users could pull down from the top of the phone's screen with their finger. Other smartphones, including the iPhone, would later adopt this feature.

Google designed the G1 to convey its philosophy about smartphones. Google wanted mobile technology to emulate the more open world of the desktop computer and Internet. Whereas the iPhone and

App Store were a "walled garden" carefully tended by Apple, users and developers would rule Android and the Android Market. "By making a lot of the source code free for others to use we knew that it would get picked up and used in ways we couldn't think of," says DeSalvo. That, in turn, would benefit Google's corporate interests as a search and advertising provider. The Open Handset Alliance and Linux were key parts of this strategy.

FROM LAGGARD TO LEADER

The ascendancy of the iPhone and Android, and the speedy adoption by U.S. carriers of a fast, fourth-generation (4G) wireless data standard, called Long Term Evolution (LTE), repositioned the United States as the center of smartphone innovation, with Silicon Valley as the new capital of the smartphone world. In 2012, the then–FCC chairman Julius Genachowski trumpeted these developments in a speech that said the United States had gone "from laggard to leader" in mobile technology.[46] He said:

> Four years ago people were talking about mobile innovation in Asia and mobile infrastructure in Europe and describing the U.S. as a backwater. Today the U.S. is the clear world leader on mobile innovation. U.S. companies invented the apps economy, and in four years the percentage of mobile devices globally with U.S.-made operating systems has grown from twenty percent to eighty percent. On mobile infrastructure, the U.S. is now leading the world in deploying at scale the next generation of wireless broadband networks, 4G LTE. . . . Today the U.S. has more LTE subscribers than the rest of the world combined, and we're on a path to maintain leadership into the future.

The changes the iPhone and Android wrought were more than geographic. Smartphones had always had more features than basic cellphones and feature phones, but after the iPhone and Android gained momentum, smartphones needed to be true mobile computers to compete effectively in the market. For many smartphone makers this shift required jettisoning their operating systems. Older platforms couldn't match Apple and Google's software prowess, which made it challenging to sell phones and entice third-party developers to write apps. These companies had to either align themselves with Google and Android or build their own mobile computing platforms. Companies that chose the latter needed to develop that software as quickly as possible while still selling and supporting phones based on the software they were phasing out. It was a tricky maneuver, and many smartphone companies fumbled the transition. Some that had led the industry in the early 2000s would cease to exist just a few years later.

One company that decided to take on Apple and Google was Microsoft. Around 2009, Windows Mobile was losing developer and licensee support in part because it couldn't match the iPhone or Android in aesthetics or ease of use. Windows Mobile didn't support multitouch, and it felt clunky on regular touchscreen phones. As the *Guardian* observed at the time, Windows Mobile "has an image problem. . . . [T]he Windows phone platform has been regarded as a dull tool for corporations instead of a strong player in the consumer market, and its user interface has never been much to write home about."[47] Or as tech reporter Dan Frommer put it, "[It] isn't that Microsoft didn't see the mobile revolution coming. It just peaked way too early, and wasn't prepared when Apple showed up with something a lot better."[48] Microsoft realized it needed to start coding a new solution from scratch rather than continue to tweak its mobile software. In 2010 it scrapped Windows Mobile and introduced more modern software. To highlight its break with its past, Microsoft rebranded its smartphone software yet again, this time calling it Windows Phone.

Meanwhile, Nokia's global reach kept Europe, Symbian, and itself at the top of the smartphone sales charts until 2011, but it started to sputter after Android's release for some of the same reasons as Windows Mobile. Symbian's code was bloated from years of additions and revisions, causing app developers to shun it. Nokia had also steered Symbian away from supporting touchscreens, which left Nokia playing catch-up once smartphones became synonymous with multitouch technology. Under a new CEO, Stephen Elop, Nokia started phasing out Symbian in 2011 and adopted Windows Phone as its smartphone operating system. The Symbian misstep cost Nokia its smartphone crown, and by 2013 Nokia had slipped to number ten in global smartphone sales.

In September 2013 Microsoft announced it would pay $7.5 billion to purchase Nokia's smartphone and cellphone business, and to license Nokia's patents and mapping services for 10 years. Nokia is now a very different enterprise, one focused on developing and selling telecommunications network infrastructure, mapping and location services, and advanced wireless and Web technologies—not phones.

Nokia's decline and Windows Mobile's weakness gave Android an opening. "Once those two platforms fizzled out, Android was the obvious choice to fill the vacuum," says DeSalvo, the former Android software engineer.

About a decade after introducing the R380, Ericsson exited the cellphone business. Over the years, its focus had shifted to selling telecom equipment and services. Designing consumer phones was a better fit for its handset partner, Sony, which acquired Ericsson's share of their joint venture in 2011 and turned Sony Ericsson into a Sony subsidiary. Sony CEO Kazuo Hirai has said he wants mobile products, including smartphones, to be one of the company's three main growth areas for the future.

Palm ceased to exist in 2011. In 2009, Palm rallied one last time, producing a well-respected operating system, webOS, and some interesting new phones. But the phones didn't sell well, and HP acquired a financially ailing Palm in 2010. Following management changes, HP killed off the Palm brand in 2011 and sold webOS to LG in 2013, which it uses in smart TVs.

Motorola split into two companies in 2011. The part of Motorola that makes two-way radios and voice and data systems for government agencies and corporations became Motorola Solutions. Motorola's consumer phone business, which had struggled for years to develop a successor to its best-selling Razr cellphone, became Motorola Mobility. Google announced its plan to acquire Motorola Mobility later that year and attempted to return Motorola to profitability for about a year and a half, but in January 2014 sold it to Chinese computer maker Lenovo for $2.91 billion.

BlackBerrys lost much of their allure once e-mail and instant messaging became standard smartphone features. In the 2010s, RIM weathered a series of major changes, starting with the appointment of a new CEO in early 2012; Lazaridis and Balsillie resigned, and Lazaridis retreated to a board role, then completely left the company in 2013 (Fregin had retired as a vice president in 2007). That year RIM also changed its name to that of its better-known brand and introduced a long-awaited operating system, BlackBerry 10, and several Black-Berry 10–based smartphones. The software won a decent amount of praise, especially for its virtual keyboard. So did RIM's first flagship BlackBerry 10 phone, the touchscreen Z10.

However, both products arrived too late to reverse BlackBerry's fortunes. BlackBerry posted dismal results for two consecutive quarters in 2013, including a $965 million loss for its fiscal second quarter. That August BlackBerry announced it was exploring "strategic alternatives," including a sale of the company.[49] Lazaridis and Fregin expressed interest in buying it, and rival technology firms offered to purchase some of its assets, such as its patent portfolio, but BlackBerry ultimately accepted a $1 billion cash infusion from a small group of its institutional investors to cover its short-term financing needs. As the *New York Times* wrote in November 2013, "[BlackBerry] will either fail miserably or be a remarkable turnaround. . . . [The $1 billion] investment . . . buys BlackBerry more time for an overhaul."[50] The investment deal ushered in yet another CEO—a seasoned enterprise-technology executive named John Chen, who has pledged to return BlackBerry to profitability by early 2016 with a renewed focus on serving companies and the government.

THE EMERGENCE OF SAMSUNG

As technologies and trends change, errors trip up the seemingly infallible, and companies exploit their rivals' weaknesses, new leaders have come to the fore in the smartphone market. One success story is

Samsung, which went from holding just a 3 percent sliver of the global smartphone market at the end of 2009 to 10 percent a year later, and continued to grow until early 2014.

Originally a food-trading business that branched into sugar and textile manufacturing, Samsung entered the electronics industry in 1969 when it established Samsung-Sanyo Electronics as a joint venture with the Japanese electronics maker Sanyo. In the 1970s, after Sanyo divested its shares in the joint venture, it became Samsung Electronics, and by the 1980s it was creating cellphones for the Korean market and harboring aspirations of becoming the world's largest electronics company.[51]

Network technologies helped Samsung establish itself as a global cellphone maker. Korea was one of the few countries outside the United States that decided to deploy CDMA for its 2G networks. Since CDMA wasn't widespread, only a handful of companies in the world produced CDMA phones in the mid-1990s, and in 1996 this lucky coincidence led Sprint to offer Samsung a $600 million contract to provide cellphones for its CDMA network.[52] The Sprint deal was Samsung's introduction to the U.S. cellphone market. Korea's manufacturing expertise, which the government began fostering in the 1970s to compensate for a lack of natural resources, also helped Samsung's ascent.

Samsung released a device widely regarded as its first smartphone, a Palm OS PDA-phone hybrid called the SPH-i300, in 2001. Over the next few years Samsung produced a rapid succession of smartphones that enabled it to experiment with software and design. In 2004, it became the first phone maker to offer handsets based on the three major smartphone platforms at the time: Palm OS, Microsoft's Windows Mobile, and Symbian. Around this time, too, Korea became a thoroughly connected society, with some of the world's highest Internet and cellphone usage rates and fastest wire-line and wireless speeds. These advances gave Samsung "a great test-bed" for mobile devices and features before exporting them, says Kim Yoo-chul, a technology reporter for the *Korea Times*.

Samsung stumbled at first with iPhone-style smartphones, but started catching up in 2009, when the Korean government lifted restrictive mobile software and wireless Internet regulations that had stymied smartphone development and adoption. In 2010, Samsung finally broke through with its Galaxy S handsets, which were thin, fast, and had global carrier support. There was a market need for an iPhone competitor, and Samsung delivered it. As a 2013 *New York Times* article made clear: "Apple, for the first time in years, is hearing footsteps. The maker of iPhones, iPads and iPods has never faced a challenger able to make a truly popular and profitable smartphone or tablet—until Samsung Electronics came along."[53]

3

The Smartphone Wars

On a chilly night in March 2013, more than 3,000 people gathered in Radio City Music Hall to watch Samsung introduce its Galaxy S4 smartphone. There was an electric feeling of anticipation inside the New York City theater as JK Shin, the head of Samsung's mobile division, emerged from behind a curtain. Striding to the middle of the hall's giant stage, Shin did his best to deliver a Steve Jobs–style pitch for Samsung's latest flagship phone. After highlighting the Galaxy S4's distinctive features, including a "touchless" interface that let users navigate the phone using hand gestures and eye tracking, Shin touted the phone as a life-changing device: "Once you spend time with the S4, you'll realize how it makes your life richer, simpler, and fuller."

To reinforce Shin's message, a troupe of actors took the stage. Over the next hour the actors, backed by a full orchestra, performed a series of skits showing off the S4's features, from its voice recognition to its health-tracking software. And besides the thousands in the hall, hundreds of people attended a "viewing party" in Times Square to watch the unveiling on gigantic billboards and more than 410,000 people tuned in to the YouTube feed to watch the event live-streamed.

Most smartphones debut at small press conferences. Apple's product launches are larger affairs, but they are still limited to reporters and analysts and held in a local convention center. Samsung unveiled its first Galaxy smartphone, the Galaxy S, at a 2010 Las Vegas trade show; the Galaxy SII in Barcelona, at a different trade show; and the Galaxy SIII at a London exhibition center. With the Galaxy S4, Samsung pulled out the stops. For Samsung, the event wasn't just about a device. It was the Korean company's chance to beat Apple on Apple's home turf. Samsung booked a famous location and hired actors, writers, and producers to put on a show in an attempt to one-up its biggest competitor.

Mixed reaction kept the Galaxy S4 launch from being a triumph. There was a near universal dislike for the skits, which were packed with embarrassing jokes. People were divided on the phone's merits as well. Nevertheless, the consumer and media buzz generated by the launch—and the hype that preceded it—cemented Samsung's position as Apple's most serious smartphone competitor. The *New York Times* called the launch "a challenge in iPhone's backyard."[1] AllThingsD (now *Re/code*) wrote, "It's usually Apple, not Samsung that gets this kind of attention."[2] *Barron's* said, "Samsung's latest shot in the smartphone wars . . . debuted with a loud bang."[3]

Selling smartphones is one of the fiercest fights in global business today. We talk about smartphone wars because the smartphone industry has become a battlefield for the world's technology giants, which scuffle to stay ahead, and sometimes just to stay in business. These companies brawl daily, on multiple fronts, both in the courtroom and in the public eye. They pour millions of dollars into lawsuits and ads attacking each other, looking for any way to gain ground. The companies themselves use warlike terminology when discussing their competitors. Steve Jobs talked about "go[ing] to thermonuclear war"[4] on Google for allegedly copying Apple's technology and also referred to engaging in a "Holy War"[5] competition with Google, on a number of fronts, including Android. Jobs's successor, Tim Cook, vowed to "use whatever weapons we have at our disposal"[6] to protect Apple's intellectual property.

The smartphone wars are intense because the market is large and lucrative. Estimates of its size range between $250 billion and $350 billion, which is larger than the PC market and more than twice as large as the Internet advertising market, although both of those markets existed years before smartphones. Smartphones also outsell PCs by a factor of more than three to one in terms of units, even though PC shipments outnumbered those of smartphones as recently as 2010. With a billion new smartphones shipped to retailers annually, market researchers say the smartphone industry is growing 19 percent each year, while growth in the PC market declines year over year. Smartphones are also outselling basic phones globally, having accounted for 55 percent of all cellphone shipments in 2013.

Smartphones aren't just a huge market; they can also be far more profitable than most other tech products. Their profit margins are high because smartphone makers can command lofty prices for their devices while reducing their production and distribution costs through large volumes. The iPhone is Apple's biggest moneymaker, accounting for more than half of the company's 2013 revenues. Analysts estimate Apple's iPhone gross product margins are between 40 percent and 50 percent.* In contrast, gross margins for Apple's other products, including the iPad, iPod, and Mac computers, hover between 20 percent and 30 percent, either because they are relatively more expensive for Apple to produce or because Apple sells popular, more affordable versions of those products, such as the iPad mini. Samsung's gross margins are lower than Apple's but show a similar gap between phones and other gadgets. Samsung's phone business is so lucrative, it accounts for two thirds of the company's total operating profit.

Naturally, smartphone companies want to protect their high profits. They also recognize that smartphones are the future of the Internet, computing, communications, advertising, and shopping. Failing in the smartphone market means losing influence in multiple important industries, and these high stakes fuel the smartphone wars.

ANDROID VERSUS APPLE

Hundreds of companies participate in some way in the smartphone economy, from contract manufacturers to gadget recyclers. The main players are the platform (operating system) providers, the smartphone makers, and the carriers. The leading carriers usually differ from country to country, but the platform providers and the major phone makers operate on a global scale. Among those global players, there

---

*Gross margin is a product's sales price less its production costs—meaning, the cost of components, raw materials, storage, and plant labor and overhead—divided by its sales price. It shows how much Apple profits from the sales of iPhones without including marketing, R&D, and administrative expenses.

are three main combatants: Apple, Google, and Samsung. Apple and Google power the most smartphones through their respective operating systems, iOS and Android, while Apple and Samsung sell the most smartphones.

Apple and Google together control 94 percent of the global smartphone market. Android holds 79 percent of the market and iOS has 15 percent. (In the United States, the split is 51 percent Android, 44 percent iOS.) More than 1.5 million Android devices, the majority of which are phones, are activated every day, while 794 million Android phones shipped to retailers in 2013. Apple sold 150 million iPhones in 2013. iOS initially led Android because it had a year's head start, having debuted in 2007 with the first iPhone, but Android passed iOS in phone shipments in mid-2010.

Apple and Google have different business priorities for their smartphones. Google, at its core, is an online advertising company. It has three main business goals: to increase the number of places it can run ads; to show those ads to as many consumers as possible; and to collect and mine consumer data to improve its ads. Android helps Google do all three. Search for a business within Google Maps on an Android phone and an ad will appear at the bottom of the screen. Google earns money by serving up this data, and it earns even more if consumers engage with the ad by, for example, clicking a link to call the business.

Android devices give Google a direct route to consumers. Since Google's Android partners preinstall its services on their phones, the operating system serves as a defense against companies that might try to inhibit Google's access to consumers on mobile devices, says Horace Dediu, a technology analyst who runs his own consultancy called Asymco. Before the iPhone launched, Google was concerned about Windows Mobile dominating the mobile industry, so that smartphones would default to Internet Explorer and other Microsoft services, leaving no room for Google's properties. After the iPhone launched, Google worried about being overly dependent on Apple for mobile traffic. Android creator Andy Rubin has said that, pre-Android, "it was very, very hard to get Google services in front of the faces of users on cellphones."[7]

Google measures Android's success by the number of users it reaches. To encourage device makers to use Android, Google makes the software high quality, free, and easily malleable. In a market as diverse and price sensitive as smartphones, this is a recipe for instant popularity. But Google also imposes more rules on device makers than outsiders might assume, given its many statements about openness. Phone makers that want to use the Android name, preinstall Gmail, Google Maps, YouTube, and other Google services on their handsets, or use Google's development tools, which support in-app features such as billing for purchases, user geolocation, and messaging, must comply with Google's compatibility requirements, which are basically tests that ensure Android apps will run smoothly on the devices. Preloading one Google service also means preloading all of them; phone makers can't take Google Maps but pass over Google+, Google's social networking service. This makes Android, in technology writer Steven Levy's words, "a gateway drug to Google products and ads."[8] Google's Open Handset Alliance phone partners further agree not to employ Android in unauthorized ways, such as "forking" Android to create noncompatible versions of the code or manufacturing devices for software companies that have forked Android. Google recently added another requirement: that new Android phones display the words, "Powered by Android," on the bottom of their screens when they are activated. Device makers that reject these conditions can use the open-source version of Android but will have to find their own replacements for Google's signature services and software tools.

One of Google's most important licensed apps is its online marketplace, Google Play, which replaced the Android Market in 2012 and sells apps, songs, books, and videos. Like other platform providers, Google takes a cut of each paid app sale, but it distributes much of its earnings to its carrier and phone maker partners, reducing the amount it pockets.

Google values Android apps because they help it sell ads, which contribute the "vast majority"[9] of its mobile revenues. Calculating what that means and the amount of money Google makes off Android specifically is tricky. Android is presumed to be profitable as a Google business division (based on advertising generated through Android).

Rubin said it was in a 2010 AllThingsD interview,[10] and Android has grown exponentially since then. But Google doesn't break out mobile ad sales in its financial results and rarely discusses them in detail during its earnings calls. The most recent occasion when Google specified its mobile revenues was in October 2012, when it said it had a "mobile run rate"[11] of $8 billion a year. Based on that number, analysts estimated the company pulled in approximately $5 billion a year from mobile ads and services across all smartphones (not just Android).[12] Google's mobile revenues have increased since October 2012, but it's not clear by how much. For example, in June 2013, Google said a new video ad format helped YouTube triple its mobile revenues since late 2012.[13] Bloomberg quoted an analyst who estimated this meant Google was reaping as much as $350 million from YouTube mobile ad revenues per quarter, but Google did not specify an amount itself.

Much of Google's mobile ad money is believed to come from iOS, since Google is the default search engine on the iPhone. Advertisers generally pay higher rates to reach iPhone users, because iPhone users are more likely to engage with mobile marketing campaigns and make purchases on their phones, according to the marketing data provider comScore.[14]

If Google measures its success through advertising, Apple's core business is selling mobile devices. Like Google Play, Apple's App Store isn't a major profit source for the company. Says Dediu, "iTunes is to Apple what Android is to Google—a business not designed to make money in and of itself but to sustain another business." The App Store is expensive to operate because Apple pays not only for the technology infrastructure, Web hosting, and payment processing but also for the army of people it employs to inspect apps and communicate with developers. "Apple considers the quality of its app offerings to be very important, so it invests a lot into app certification, which is very costly," says Michael Vakulenko, strategy director at the London-based market analysis firm VisionMobile. VisionMobile has estimated that, on average, it costs Apple about $2,500 to put an app in the App Store, regardless of whether it is a free or paid app.

The flagship iPhone is Apple's profit engine. Apple is believed to pocket more than $200 in profit from each one, due to efficient manufacturing and high sales prices. Apple also leverages its size and cachet to squeeze low prices from component makers and other suppliers. It also rigorously limits the number of iPhone models it sells. By mostly restricting its smartphone offerings to its three most recent models—for instance, the flagship iPhone 5S, the cheaper iPhone 5C, and the older iPhone 4S—it can exploit economies of scale in sourcing and manufacturing.

Sales to carriers are the other half of the iPhone profit formula. Though other smartphone makers charge carriers a discount wholesale rate for their phones, Apple demands full price. Carriers pay about $650 for every flagship iPhone they purchase. Apple also forces carriers to buy a certain number of iPhones in a given time period. For example, in 2011, Sprint agreed to purchase $15.5 billion worth of iPhones over four years in exchange for the right to finally sell the device.[15] Carriers loathe Apple's tough terms, but most acquiesce to its conditions because the iPhone is incredibly popular with consumers, and iPhone users buy expensive mobile broadband plans, which ultimately help the carriers.

Carrier support has helped keep the iPhone's price sky-high for years—a rare occurrence in an industry where smartphones generally experience dramatic price drops a few months after launch. By 2011, Apple was so flush with iPhone and iPad profits, it passed oil and gas company ExxonMobil to become the world's most valuable company by market capitalization. ExxonMobil retook the title in the spring of 2013, but Apple has since managed to reclaim the market-cap crown, in part due to ExxonMobil's own challenges, including higher production costs for oil and natural gas.

In 2013, Apple introduced the iPhone 5C, which was $100 more affordable than previous iPhones, and in 2014, it launched a version of the handset that was $70 cheaper, with a smaller amount of internal memory, in major markets outside the United States. But Vakulenko doesn't think Apple will unveil a truly low-cost iPhone for a few years, because Apple prizes product quality and profitability above market share. As Tim Cook likes to say, "We have never been about selling the most."[16] Instead, Apple creates value for iPhone users by maintaining an ecosystem that connects developers to consumers. Vakulenko says that as long as that ecosystem is large enough to sustain itself and be competitive with that of Google's Android, Apple won't worry about the need to introduce a low-cost iPhone. "It's a delicate balance," Vakulenko explains. "Apple is trying to optimize profit, so it will go as low as it thinks it can go [price-wise] to satisfy the needs of its [ecosystem], but it will also stay as high as possible to keep the company hugely profitable."

ECOSYSTEM WARS

Most people think of the smartphone wars as individual companies and phones competing head to head, but the industry is also a war of ecosystems. Apple and Google both operate self-reinforcing ecosystems based around apps. These ecosystems helped Apple and Google grab power within the smartphone industry several years ago, and now help them preserve their power.

The iOS and Android ecosystems function essentially the same way. They link two groups that normally have difficulty communicating: developers and consumers/users. The groups feed each other, and the resulting sales feed Apple and Google. More precisely, the large number of iOS/Android apps attract users. Those users attract new developers and encourage them to produce apps. Those apps then attract new users to iOS/Android. It's a positive feedback cycle that benefits the entire ecosystem, particularly the ecosystem owners, Apple and Google.*

Vakulenko says these "network effects" are found in all computing systems. Computing companies have welcomed developers and third-party apps since the 1980s. As a computer maker, Apple understood the strength of this model though Steve Jobs initially resisted the idea of opening up the iPhone. As a provider of cloud computing and Internet-based services, Google also understood it. Smartphone makers without computing experience, such as Korea's LG and Motorola, grasped the concept late. Other phone makers established eco-

---

*This discussion concerns official ecosystems. There are also a number of unofficial Android ecosystems.

systems relatively early, but they weren't app-based. Vakulenko says the BlackBerry ecosystem mostly connected BlackBerry users to each other through messaging technology, and Nokia's Symbian ecosystem mostly served the needs of phone makers and carriers. Since those ecosystems didn't foster the development and distribution of apps, they didn't catch on like iOS and Android did.

Successful smartphone ecosystems flourish for years. Network effects enable these ecosystems to grow exponentially. Strong ecosystems, such as iOS and Android, further lock in users by selling them smartphone accessories. People who spend money within a smartphone ecosystem are less likely to defect to competing platforms, because they don't want to give up their purchases. Habits are also hard to break, and people build up comfort and familiarity with the way their smartphone platforms are designed and operated. To lure these consumers away a new ecosystem has to match those apps and accessories as well as offer something uniquely appealing.

Smartphone ecosystems also lock in developers by convincing them to invest considerable time and effort learning their specific coding languages and business systems. "There's a complex set of issues developers need to master on a new platform, from how to use [software tools] to how to monetize their apps," says Vakulenko. "Once developers get proficient, they need a very, very good reason to jump somewhere else and learn again from scratch." The most popular ecosystems, such as Android and iOS, also provide developers with useful tools, such as ways to translate their apps into foreign languages. In smaller ecosystems developers have to handle these tasks themselves or forgo them entirely.

These high switching costs have kept users and developers loyal, forging iOS and Android into smartphone empires that can ward off upstart ecosystems that try to replace them. So far, only one newer ecosystem has managed to gain a foothold, and it's a small one. Windows Phone is the industry's distant number three, with about 3 percent of global smartphone sales.

Windows Phone faces an uphill battle. Timing is a crucial factor in the smartphone wars, and Windows Phone was a few years behind the competition. By the time Microsoft phased out Windows Mobile and brought Windows Phone to market, it was late 2010. iOS and Android had already rounded up and locked in droves of consumers and developers. Also, until early 2014 Microsoft monetized Windows Phone through licenses to phone manufacturers, reportedly charging companies including HTC and Samsung somewhere between $5 and $15 for every Windows Phone handset they sold. This business model put Microsoft at a disadvantage in an Android-dominated world. "Smartphone vendors are already squeezed so much trying to get costs down," says Dediu. "On top of that, you have Microsoft saying you have to pay this license. The alternative is Google, who says, 'You don't have to pay for Android, which is just as good, if not better.' "

Windows Phone managed to pick up momentum anyway, in part because of RIM/BlackBerry's decline. Windows Phone was the world's fastest-growing smartphone operating system in 2013 (though it also started from a much smaller base), and it accounted for more than 10 percent of the smartphone market in some European countries by early 2014, thanks to Nokia's historic strength in Europe. Nonetheless, in April 2014, Microsoft took the dramatic step of making Windows Phone free for smartphone makers. Microsoft appeared to realize that nothing short of free distribution would enable it to grow the Windows Phone ecosystem beyond itself and Nokia, whose cellphone and smartphone business it was on the brink of absorbing.

The big question is how Microsoft plans to make money off Windows Phone now that it is giving it away. Microsoft's new CEO, Satya Nadella, who succeeded Steve Ballmer in February 2014, frequently talks about how he is directing the company to embrace a "mobile-first cloud-first" world. Vakulenko approves of that plan, as it shifts Microsoft's focus from "the stagnating PC business" to "high-growth areas." PC shipments have been sliding since mid-2012 and posted their largest-ever annual decline (of 10 percent) in 2013. But Vakulenko maintains that "it is not clear how exactly Microsoft will . . . build a profitable business in these areas." He notes that successful smartphone platform providers either use services to increase the value of their devices (Apple) or commoditize devices to achieve the widest distribution possible of their services (Google). Microsoft is pursuing a different strategy that uses devices to sell cloud-computing services, such as its online Office 365 service. Nadella has said the company's new mission is to "deliver the best cloud-connected experience on every device."[17]

Even if Microsoft fails to establish Windows Phone as a major growth market, it will probably continue to fund Windows Phone to bolster its ecosystem. In a 2013 PowerPoint presentation it posted online that outlined its "strategic rationale" for acquiring Nokia, Microsoft wrote: "Success in phones is important to success in tablets" and "Success in tablets will help PCs."[18] In other words, Microsoft believes a person who buys a Windows Phone is more likely to buy a Windows tablet and a Windows computer, especially with its recent introduction of "universal" apps that will work across all Windows devices. Tim Cook has said this is true for Apple, stating in 2012, "What is clearly happening now is that the iPhone is creating a halo for the Macintosh [computer]. The iPhone has also created a halo for iPad."[19] In other words, the smartphone wars aren't just about phones. They're also about full ecosystems of devices and services.

APPLE VERSUS SAMSUNG

Like smartphone platforms, smartphone manufacturing is a two-horse race. Here, instead of Apple and Google, the two are Apple and Samsung, and they are so far ahead of other phone makers that they currently claim more than 100 percent of global smartphone profits. In the fourth quarter of 2013, Apple took 76 percent of smartphones profits while Samsung took 32 percent, according to the investment bank Canaccord Genuity. Apple and Samsung were able to amass 108 percent of the profits because other smartphone makers lost money or just broke even.

While Apple leads the industry in profitability, Samsung leads in volume. Samsung is the world's largest smartphone maker by unit sales, selling about 31 percent compared to Apple's 15 percent. Outside of Samsung and Apple, the market splinters. No other company holds more than a 5 percent share.

Analysts often describe Samsung as a "fast follower" because it sweeps into a market after other companies have laid the groundwork.

It has done this for decades, with one book about the company describing its strategy as follows: "Samsung's main thrust boiled down to picking specific areas to achieve key differentiation, the most successful of which was the mobile cellular phone. Samsung then focused on flooding the market."[20]

Even now Samsung floods the market. It has the largest product portfolio of any major smartphone maker, enabling it to attack Apple at the high end of the market and Nokia at the low end. "The Samsung business model is about muscle, not finesse," says Dediu. "Companies like Apple establish a beachhead and Samsung expands it, then gets out before there's nothing interesting left." In the United States alone Samsung offers close to 50 smartphones at prices ranging from $300 to free with a two-year contract. Globally Samsung sells more than 70 different smartphones across three operating systems. "While Apple has [mostly] been sticking to its 'release one model a year' philosophy, Samsung took a different path, expanding its lineup from premium to mid- to low-range smartphones," says the *Korea Times*'s Kim Yoochul. Many Samsung smartphones are unremarkable, but Samsung has turned its high-end Galaxy S and larger-sized Galaxy Note into household-name franchises that together accounted for more than 100 million shipments in 2013.

Flooding the market isn't a winning strategy by itself. What sets Samsung apart is its more integrated approach toward production. In smartphone manufacturing, integration is a major advantage, because it allows companies to move quickly and wield more control over phone appearance and performance. Apple is able to integrate its hardware and software because it designs the iPhone, its operating system, its core services, and some of the phone's most important components, such as the iPhone's processor. Samsung is able to integrate its phone hardware down to the component level because it makes many of its phones' parts, including screens, processors, and memory chips. These particular components attract high consumer interest and are some of the most expensive parts, which has been a boon to Samsung.

Most smartphone makers outsource the creation of their components and operating systems to other companies. Apple and Samsung's synergies enable them to be first to market with the most cutting-edge smartphones. Being first is important, because carriers and retailers are always looking for new phones to promote and consumers are always looking for innovative phones to buy. In this hypercompetitive industry, speed and control translate into sales and profits.

Tight integration and comprehensiveness aren't Samsung's only weapons: it also uses marketing to outflank its competitors. In 2012, Samsung spent more on global marketing ($11.6 billion) than on R&D ($10.3 billion). More than $400 million of that money went to the United States for ads that feature its phones. Apple spent 17 percent less ($333 million) on U.S. ads for the iPhone that year. In 2013, Apple increased its spending 5 percent to about $351 million, and Samsung reduced its by 10 percent, but still spent more—about $363 million.[21] In 2014, Samsung paid an estimated $18 million to run commercials during that year's Academy Awards broadcast—a deal that resulted in host Ellen DeGeneres using a Galaxy Note 3 to organize a star-studded "selfie" photo that, when pushed to Twitter, became the most retweeted tweet ever.[22] Reuters has called the company "the world's biggest advertiser."[23]

Samsung also invests in co-marketing, which are payments phone makers make to carriers and other retailers to promote their phones. (In contrast, Apple expects carriers to devote much of their marketing budgets to iPhone ads to meet the sales obligations in their iPhone contracts.)[24] Carriers use co-marketing funds to run TV, online, outdoor, and in-store ads about devices and to offer special promotions. For example, when a carrier gives consumers a free Samsung phone with the purchase of another Samsung smartphone, and heavily advertises the offer, Samsung is probably paying some portion of those expenses. Smartphone makers say Samsung gives carriers more co-marketing money than other companies do, which makes sense given its size and profitability. The more money Samsung gives carriers, the more ads and promotions they run featuring Samsung phones.

Samsung's next target is the big smartphone ecosystems. As a phone maker, Samsung doesn't need to establish an app ecosystem; it already participates in those of Android and Windows Phone through its devices. But having its own ecosystem would give Samsung more control over its products and yield more profit. As Dediu has written, "Samsung must either stretch into becoming one of the ecosystem contenders or be relegated to a commodity hardware company."[25]

Samsung has been trying to build an ecosystem since 2009, first by introducing its own smartphone operating system bada, which it used outside the United States and has since retired, and now by backing the open-source, Linux-based Tizen, which also has support from the chip maker Intel and six Asian and European carriers. In March 2013, Aapo Markkanen, a senior analyst for the market research firm ABI Research, wrote, "All signs are pointing to Samsung trying to pull off a Great OS Escape [from Android to Tizen] within the next year or two."[26] Samsung prepared for this escape by creating its own Android-based apps that duplicated some of the functionality of Google's. Samsung also spent freely to shore up its weakness in software. Internally, Samsung directed half of the employees in its research and development division to work exclusively on software projects. Externally, Samsung hosted its first-ever developer conference in San Francisco in October 2013, followed by a gathering for European developers in London. The goal: to recruit developers to make apps specifically for its devices, first on Android and later on Tizen. Samsung also went on a Silicon Valley hiring and construction spree. In less than a year it established a start-up accelerator and a strategy and innovation center and broke ground on a new campus for R&D and sales staff.

The moves raised Samsung's profile among American start-ups, software firms, and developers, but in January 2014 Google brought Samsung back into the Android fold by selling Motorola to China's Lenovo and signing a 10-year patent cross-licensing agreement that lets Google and Samsung use technologies covered by the others' patents. Since Samsung and other Android device makers regarded Google as a competitor after it acquired Motorola and began producing smartphones, shedding Motorola improved Samsung-Google relations. At the same time, Google saw a benefit in combining Lenovo and Motorola, because they would create a strong number-two Android phone vendor that could keep Samsung in check. Vakulenko says Android's success depends on having a "balance of power" among its phone makers, and Google sold Motorola to restore that balance.

While Samsung hasn't made its Great OS Escape yet, it doesn't appear to have scrapped the idea, either. It continues to develop Tizen and is using it as the operating system for its Gear 2 smartwatches, which wirelessly connect to its Galaxy smartphones. Samsung executives have said that Android "still needs to be our main business," but that Tizen will be an important secondary platform for the company.[27]

Samsung's rise has already touched off a war with Apple. Samsung used to produce several iPhone parts, including the processor and screen; the very first iPhone contained a Samsung processor and memory chip. But when Samsung started challenging Apple's smartphone sales, Apple dropped it as a key iPhone supplier. In 2010, Apple hired Sharp and Toshiba to make iPhone screens.* In 2013, the *Wall Street Journal* reported that Apple had tapped Taiwan Semiconductor Manufacturing Company (TSMC) to make iPhone processors.[28] Apple's new suppliers are leaders in their fields, but analysts say Apple primarily switched to slash Samsung's profits and keep its iPhone technology plans and business forecasts confidential from its fiercest competitor. According to the *Wall Street Journal*, TSMC took several years to produce chips that met Apple's high standards, while Sharp and Toshiba had to expand their factories just for Apple.[29] In contrast, Samsung had already mastered these processes.

Samsung is anticipated to lose at least $10 billion in annual revenues if Apple fully backs away. That could happen in 2015, although some experts, including the *Korea Times*'s Kim Yoo-chul, don't believe Apple will completely drop it. Kim notes that Samsung is known to guarantee on-time delivery, high levels of production, and competitive pricing to its customers, which makes it an attractive manufacturing partner. "Although it's true that Apple is cutting its reliance on Samsung, that doesn't mean Samsung will be excluded from the development of A-series [iPhone processor] chips," contends Kim. "Apple is ordering some chips from TSMC as part of its strategy to

---

*Toshiba's mobile display business is now part of a joint venture called Japan Display.

diversify its sourcing, but Samsung will remain a top-tier Apple supplier." Recent media reports in Korea and Taiwan[30] indicate Samsung
may retain 30 percent to 40 percent of Apple's chip business.

PATENT WARS

Patents are another battleground in the smartphone wars. The current
spate of lawsuits kicked off in 2009 when Nokia sued Apple for allegedly violating its wireless technology patents. Since then, every major
smartphone maker and platform provider has sued or been sued by
at least one competitor. This not so exclusive group includes Apple,
BlackBerry, Google, HTC, LG, Microsoft, Motorola, Nokia, Samsung,
and Sony. Of them, Apple and Microsoft have been the most proactive
about protecting their intellectual property (IP) rights—or, from
some people's perspective, the most ruthless in attacking their rivals.

Smartphone patent lawsuits generally allege infringement of technical inventions or designs or both, and the plaintiffs aim to block
competitors' sales, force them to alter their phone designs or technology, or elicit large lump sums or ongoing fees. Since patents can be
brandished both offensively (to sue another company) and defensively
(to ward off lawsuits), the patent fights have fueled billions of dollars of patent purchases. David E. Martin, the chairman of intellectual
property management firm M-CAM, describes this frenzied accumu-
lation of patents as "the aggregation of warheads to use for litigation."

In 2010, Apple and Microsoft teamed up with a few other tech
companies to buy 882 patents mostly related to open-source software for $450 million from the computer software firm Novell. (Cit-
ing antitrust concerns, the U.S. Department of Justice later ordered
Microsoft to sell back its share of the Novell patents and license them
instead.) The following year, a team that included Apple, BlackBerry,
Microsoft, and Sony won the biggest-ever auction of technology patents by spending $4.5 billion to buy 6,000 patents and patent applica-
tions from Nortel, a bankrupt telecom equipment provider. Nortel's
patent portfolio covered the full breadth of telecommunications, including voice communications, data networking, semiconductors,
and—most valuable of all to smartphone companies—wireless tech-

nologies, including high-speed 4G LTE. Google also bid for the Nortel patents, but it lost. To compensate, Google purchased 2,000 cellular, online search, and wireless technology patents from IBM for an undisclosed sum and spent $12.5 billion to buy Motorola, its largest-ever acquisition.

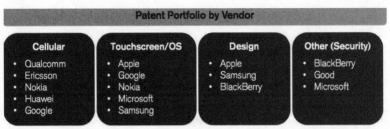

The main types of smartphone patents and the companies that hold them (*Kulbinder Garcha/Credit Suisse*)

Due to their pricy Nortel and Motorola purchases, Apple and Google spent more on patents than on research and development in 2011. Google has said Motorola's more than 17,000 patents are worth $5.5 billion, but during the years Google owned Motorola, it was not able to assert them against its competitors in a significant way.* Martin says that what Google really bought was "annoyance value." "The number of patents is valuable even if the patents themselves are worthless, because [opposing] lawyers then have to read a lot of patents," he explains.

Much of the public attention on smartphone patent fights has focused on these billion-dollar purchases, new lawsuit filings, outsize jury verdicts, and companies' attempts to crimp each other's sales and imports through injunctions. However, Florian Mueller, a noted IP analyst who writes the popular blog FOSS Patents and consults on wireless devices for companies, including Microsoft, says the battle is more complex than the typical winner-take-all narrative. "This is

---

*Google will retain ownership of the "vast majority" of these patents after the Motorola sale and transfer of about 2,000 to Lenovo, along with a license that lets it utilize the remainder.

a much more sophisticated game than a boxing match where you go into the ring for twelve rounds—or less, in the event of a K.O.—and have a result after about an hour," says Mueller. And landing a quick knockout is tough in smartphone patent fights, which often pit deep-pocketed giants against other deep-pocketed giants. Says Mueller, "The U.S. legal system is simply very slow if there's a well-funded, sophisticated defendant prepared to exhaust all procedural options and to stall."

Consider the multiyear, ongoing feud between Apple and Samsung. Apple started the skirmish in April 2011 with a 373-page complaint that alleged Samsung had knocked off its iPhone and iPad software features and designs, that it needed to halt its infringing behavior, and that it owed Apple monetary damages. According to Apple's filing:

> Instead of pursuing independent product development, Samsung has chosen to slavishly copy Apple's innovative technology, distinctive user interfaces, and elegant and distinctive product and packaging design, in violation of Apple's valuable intellectual property rights. . . . Samsung has made its Galaxy phones and computer tablet work and look like Apple's products through widespread patent and trade dress infringement. Samsung has even misappropriated Apple's distinctive product packaging. . . . Apple seeks to put a stop to Samsung's illegal conduct and obtain compensation for the violations that have occurred thus far.[31]

The two companies have since sued and countersued each other in ten countries across Asia, Australia, Europe, and North America in what Apple referred to as "a worldwide constellation of litigation"[32] in its filings. Both companies have offered to license their patents to the other, but each rejected the proposed rate as too high. Apple also wanted any proposal to include an "anti-cloning" provision that would let it sue Samsung in the future for products it deemed overly similar to its own, according to Mueller.

The most notable Apple-Samsung fight took place in a U.S. district court in San Jose, California. In August 2012, following a widely chronicled four-week trial that the *Wall Street Journal* dubbed THE

Patent Trial of the Century,[33] a San Jose jury decided Samsung had infringed six Apple patents—three utility (technical) ones and three of design—and owed Apple $1.05 billion in damages. The utility patents covered: so-called rubber band technology, which gives a bounce-back effect when smartphone users scroll beyond the top or bottom edge of a page; "pinch to zoom" multitouch technology, which can distinguish between a vertical/horizontal finger movement and a diagonal movement,[34] enabling recognition of pinch and reverse pinch movements; and "tap to zoom and navigate" technology, which refers to users navigating to another part of the phone screen after they tap their phone screen twice to minimize or enlarge text and images. The design patents covered the iPhone's appearance, including the distinctive metal rim that surrounds its glass display and the grid of round-cornered icons on its home screen. The more than 20 implicated Samsung phones included the original Galaxy S and SII, as well as other Android phones released in 2010 and 2011.

Six months later, citing jury errors, the judge vacated the damage fees associated with 14 of the Samsung products (later amended to 13) and scheduled another trial to recalculate them. The decision from that trial, which took place in November 2013, set the amount Samsung owed Apple at $290 million, which changed the total amount of payable damages to about $930 million. Since the revised award was about 90 percent of the amount initially determined by the 2012 jury, it was interpreted as a disappointment for Samsung. As Mueller wrote, "Samsung had challenged last year's jury award but obviously intended to get more out of a retrial than a 10 percent discount."[35]

In 2014, Apple and Samsung faced off again in the same court over a different set of utility patents* and mostly newer devices, including

---

*The three Apple patents that Samsung was found to infringe covered: a "slide to unlock" feature that enables users to unlock their phones by moving an image across the screen; a particular "auto-correct" method of displaying suggested words when a user is typing; "quick links," i.e., a system that detects data, such as phone numbers and addresses within e-mails and text messages, and lets users swiftly take actions such as calling the number or adding the information to a calendar. The one Samsung patent that Apple was found to infringe covered methods of displaying and classifying photos and videos.

the Galaxy SIII, the Galaxy Note and Note II, and the iPhone 4, 4.5, and 5. Apple originally sought $2.2 billion in damages while Samsung asked for $6.2 million. The jury delivered a mixed verdict, ruling that Samsung should pay $119.6 million for infringing three of Apple's five patent claims-in-suit and Apple should pay Samsung $158,400 for infringing one of Samsung's two patent claims-in-suit.

The Apple-Samsung patent battle isn't over yet. Samsung has appealed the 2012 verdict and Apple has filed a cross-appeal. Samsung has also said it plans to appeal the 2014 verdict. Smartphone patent fights can spend years winding through appeals courts. "With a very few and negligible exceptions, each and every decision by a first-instance court in these [smartphone patent] disputes has been appealed, and most of these appeals haven't been resolved yet," says Mueller. "Or if they have been resolved, the cases themselves are still ongoing, because the appeals court remanded them . . . for further proceedings." With billions at stake, neither Apple nor Samsung will capitulate easily.

Many experts view the Apple-Samsung war as an attempt to hold back the competitor. "Apple and Samsung are interested in interrupting each other's ease of market access, not in defending their patents," says Martin. "From their perspective, the more they can interrupt each other, the better." The suits can also be seen as a proxy for Apple and Google's long-simmering animosity. All the Samsung devices implicated in the Apple-Samsung suits run on Android. Apple has said Android "provides much of the accused functionality"[36] at issue in its Samsung suits. Steve Jobs famously called Android a "stolen product" and vowed to "destroy" it.[37]

However, experts such as Mueller also contend that it would be financially irresponsible for Apple to ignore allegedly infringing behavior. Since Apple's design and brand enable it to command huge price premiums, it must "defend its uniqueness" in order to protect its core business, Mueller says. "Non-enforcement is not an option for major right holders that own valuable IP" such as Apple and Microsoft, he adds.

These companies target Android device makers instead of Google because of Android's dispersed business structure. "[As a plaintiff] you don't start an outright war with [an] entire ecosystem; you focus on one particular infringer," points out Mueller. Arguing that Samsung sold infringing devices worth $3.5 billion, as Apple did in its 2013 retrial, is "a much better story" to present to a jury, he adds, than arguing that Google benefited indirectly, through advertising revenues, from patent-infringing Android devices it did not make.

Even though Google is generally not named in its Android device makers' lawsuits, it helps its partners defend themselves, to some extent. In 2011, Google transferred nine of its user interface and wireless communications patents to HTC so HTC could assert them against Apple, which it was battling at the time over an array of user interface and technical patents. Google has participated in a few Android lawsuits as an intervenor—a third party added to ongoing litigation—though it appears to intervene only when its smaller partners, such as HTC, are involved. In those cases, Google supported its partners by defending the Android-specific parts of the lawsuits. It also makes available lawyers from Quinn Emanuel Urquhart & Sullivan, its go-to law firm for patent and copyright cases. Quinn Emanuel lawyers have appeared in court on behalf of HTC (in an earlier skirmish with Apple) and Samsung (in its various Apple trials). These arrangements appear to stem from indemnification agreements between Google and Android device makers. A 2013 CNET article summed up Google's assistance to Samsung as: "quietly lending support, coordinating with Samsung over legal strategies, providing advice, doing extra legwork, and searching for prior evidence."[38] Deposition testimony from the 2014 Apple-Samsung trial further revealed that Google also offered to handle its partners' litigation and defense and to pay at least some legal costs and potential damages for patents that involve core Android features.

Google was a more active combatant through Motorola, which has tried to stymie Apple with patent claims. In 2012, Motorola (which by then was part of Google) sued Apple over seven patents related to its Siri voice-activated personal assistant and other iPhone features.

Motorola later dropped the suit. Google also inherited an ongoing Motorola-Apple infringement dispute that began in 2010, which now involves eight wireless communication and smartphone patents.

Much of Microsoft's smartphone patent activity is aimed at Android manufacturers. Claiming Android built on top of its computer operating-system and web browser software inventions, Microsoft has convinced more than 20 device makers, including HTC, LG, and Samsung, to pay licensing fees to it. "Microsoft's model has been less court action, more a quiet word in the ear," observed the Register, a British technology news site.[39] Microsoft has said the agreements cover 80 percent of Android phones sold in the United States and the majority of Android phones sold worldwide. The fees are believed to run $5 to $10 per handset and are estimated to net Microsoft $1.6 billion a year, according to global investment bank Nomura Securities.[40] If the estimate is correct, Microsoft's Android licensing business was more than four times more profitable than its Windows Phone licensing business was during Microsoft's fiscal year 2013—and thus may have helped Microsoft decide to make Windows Phone free in 2014.

Microsoft and its supporters say it is simply asserting its intellectual property rights, but Android royalties are also a way for it to obstruct Google without suing it directly. Attaching a cost to using Android reduces the software's allure to phone makers while making Windows Phone more appealing. In 2011, Google publicly expressed its ire over this "tax for . . . dubious patents."[41] The outburst didn't daunt Microsoft, which continues to strike Android licensing deals.

Besides tying up courts around the world, smartphone patent disputes have also ratcheted up activity at the U.S. International Trade Commission (ITC). Apple, Microsoft, Motorola, Nokia, and Samsung have all taken cases to the ITC in recent years. Companies petition the ITC because it acts quickly and has the power to bar imports of patent-infringing products. For cash-rich companies such as Apple and Samsung, injunctions (and subsequent design changes) are the real goal, not onetime damages payouts. Since most U.S. smartphones are assembled in China and must be imported, an ITC injunction could maim any smartphone company with a sizable U.S. business. Freezing a company's shipments at the border would allow its competitors to surge ahead, at least until the company devised non-infringing modifications known as workarounds.

ITC orders aren't always enforced or upheld, however, and presidential administrations can veto ITC bans. For decades no president exercised that power, but in 2013 the Obama administration, acting through the U.S. trade representative (USTR), overturned an ITC ban Samsung had won that affected older Apple products, such as the iPhone 3GS and 4.

Google's latest problem is a small, Ottawa-based company. Rockstar describes itself as an IP licensing company, but reporters and people in the technology industry call it a nonpracticing entity (NPE) or a patent-assertion entity (PAE). NPEs, often derided as "patent trolls," primarily buy patents for litigation and licensing purposes rather than to produce a product or service. In 2013, Rockstar simultaneously filed lawsuits against eight Android companies, including Google, HTC, LG, and Samsung. The Google suit alleged infringement of seven search advertising patents, while the other suits alleged all the manufacturers' Android devices infringed seven (and in some cases just six) patents related to mobile messaging, user interface design, and data networking. Rockstar later added Google as a co-defendant to its Samsung/Android suit. Most of the Android manufacturers have filed motions to dismiss Rockstar's suits.

NPEs are, by nature, opportunistic in their business dealings, but Rockstar aggravates Google and Samsung more than other NPEs, because it has links to Apple, BlackBerry, and Microsoft. After the Nortel auction, Apple, BlackBerry, Microsoft, Sony, and other partners divvied up about 2,000 of the 6,000 Nortel patents and established Rockstar as a company to make money off those remaining. Although Rockstar is independent from Apple, BlackBerry, and Microsoft, and its CEO insists it selects its targets itself, the company has a direct and financial connection to Android's biggest rivals. The tech news site Ars Technica called the Rockstar suits "an all-out patent attack on Google and Android" that drew upon "the patent equivalent of a nuclear stockpile."[42]

Since most companies would rather avoid the distraction and expense of going to trial, NPEs are able to extract payments and licensing agreements from at least some of the companies they threaten. NTP, a Virginia-based patent holding company, is the smartphone industry's most notorious example of NPE success. In 2006, RIM (BlackBerry), which was facing an injunction that would have seriously harmed its smartphone business, paid NTP $612.5 million to resolve a five-year dispute over wireless e-mail patents. Emboldened by its victory, NTP asserted its wireless e-mail patents in suits against other major smartphone companies, including carriers (AT&T, Sprint, T-Mobile, and Verizon Wireless), phone makers (Apple, HTC, LG, Motorola, Samsung), and platform providers (Google and Microsoft). In 2012, NTP reconciled with all of these companies for undisclosed terms.

The Obama administration tried to mitigate patent trolling with the America Invents Act, which was signed in 2011. Though it was billed as landmark legislation, the patent reform law primarily made administrative changes, such as updating patent-filing procedures, and experts say patents remain overly broad and easy to abuse. Congress is also trying to stamp out trolling with the Innovation Act, which is specifically designed to amend and improve the America Invents Act. The act includes patent reform provisions, such as requiring patent complaints to include more details about the alleged infringement and helping defendants recoup their legal fees if they prevail against a troll in court. The U.S. House of Representatives passed the bill in December 2013.

The Innovation Act won't fix one of the American patent system's chief flaws: the surplus of shoddy patents, especially software patents. (Recent attempts by the U.S. Patent and Trademark Office to improve technical training for its patent examiners and judges could help, however.) "There are too many patents being issued on stuff that isn't a true invention," says Martin. "Getting a patent [these days] is just evidence you argued with a patent examiner until he got tired and gave you a patent." By M-CAM's count there are about 210,000 cell phone patents in the world, approximately 180,000 of which relate to smartphones.[43] As of the end of 2013, Apple held 852 smartphone-related patents and Samsung had 1,964. "In a world of 200,000 mobile phone patents, what's innovation?" asks Martin. "The universe is so cluttered with patents, no one really knows what they have."

CARRIER WARS

Carriers don't generate headlines the way Apple, Google, and Samsung do. Nevertheless, the smartphone-related carrier wars should not be overlooked. Carriers fight with each other over lucrative smartphone consumers, and they fight with smartphone platform providers over industry revenues and control. U.S. carriers have their own duopoly—AT&T and Verizon Wireless, which control about two thirds of the U.S. wireless market. The number three and four largest U.S. carriers—Sprint and T-Mobile US—are employing increasingly combative and creative tactics to steal market share from the two leaders. These carrier-initiated battles shape the smartphone industry, too.

The intracarrier battles are familiar to most consumers because the tussles are highly public and are often provoked and propelled by marketing claims. Carriers pour billions of dollars into ads that attack their rivals and try to persuade consumers to switch service providers. According to *Advertising Age*, AT&T spent $2.91 billion on marketing in 2012, and Verizon was right behind AT&T with $2.38 billion.[44]

One of the most famous carrier attack ads was Verizon's 2009 "There's a Map for That" campaign against AT&T, which contrasted U.S. maps showing Verizon's nationwide 3G network coverage with ones depicting that of AT&T. Unsurprisingly, the maps made Verizon's network look far superior. AT&T sued Verizon for "misleading" consumers, which it said was causing it to lose "incalculable market share,"[45] and it ran retaliatory ads in an attempt to refute Verizon's allegations. By the time AT&T dropped the suit a month later, the clash had attracted mass attention. *Advertising Age* called it "a big-bucks marketing battle that's already threatening to make the cola wars [between Coke and Pepsi] look like child's play."[46]

In 2013, Verizon renewed its map marketing attack with a series of commercials that used visual comparisons of maps—hung in a gallery like abstract art, in one ad—to show Verizon covered more of the country with 4G LTE than AT&T, Sprint, or T-Mobile. But these days the fiercest advertising war is between AT&T and T-Mobile. Recently appointed T-Mobile CEO John Legere touched off the quarrel in early 2013 when he called AT&T's network "crap" at the Consumer Electronics Show (CES), the world's largest consumer technology trade show.[47] AT&T and T-Mobile have traded insults ever since, in both newspaper and TV ads. AT&T has said T-Mobile has twice the number of dropped calls and half the download speed as its own network. T-Mobile has accused AT&T of deliberately confusing and overcharging its customers, and Legere regularly slams AT&T on his Twitter account; he even crashed AT&T's 2014 CES party (only to be escorted out by security). In 2014, both carriers offered to pay the other's subscribers hundreds of dollars to switch to their networks. (T-Mobile's "break up with your carrier" campaign also extended to Sprint and Verizon customers. Sprint later launched a similar promotion.) Analysts say T-Mobile's aggressive marketing is helping it gain subscribers, so the carrier will probably continue hurling insults for a while.

Carrier animosity toward Apple and Google is subtler—more a case of simmering tension than in-your-face smack talk. Carriers resent the way platform providers usurped their power over smartphone subscribers and their revenues from mobile content sales. Before the iPhone launched, carriers controlled the smartphone ecosystem. Everything a user did, whether it was talking, texting, or accessing the Internet, took place on the carrier's network, and the carrier got paid for it. Carriers also operated online portals, where they sold mobile content such as ringtones and games, steering their subscribers to these portals and pocketing most of the profit from content sales. Analysts often described this setup as a walled garden, similar to Apple's App Store, but with carriers as the gatekeepers.

The iPhone set off a chain of events that shifted the smartphone industry's balance of power. As the mobile industry transitioned from mobile telephony to mobile computing, platform providers and smartphone makers pushed carriers to the sidelines. Apple first loosened the bonds between carriers and subscribers. Before the iPhone, smartphone makers didn't have close ties with their users; they let carriers fill that role. Apple co-opted the carriers' customer relationships by selling the iPhone directly to consumers and designing it to be activated, registered, and updated through iTunes. iPhone owners started looking to Apple for information about their phones.

Apple also poached the carriers' mobile content business by opening the App Store. Money from apps and mobile media that used to flow to the carriers through their online portals now goes to Apple, Google, Microsoft, and their platform-specific app stores. Google mollifies carriers by giving them part of the 30 percent cut it gets on app sales. "Google paid operators [carriers] because it wanted to flood the market with Android devices," says Vakulenko. "It needed to attract operators' support and make sure they were putting marketing into promoting Android." But now that Android is a runaway hit, Google appears to have changed its policy. It is reportedly trying to renegotiate its carrier agreements to give itself a bigger percentage of app revenues.[48]

These developments threaten to make carriers into generic providers of utility services. In industry lingo, the phrases are "dumb pipes" or "bit pipes," as in the data bits and bytes carriers transfer between their subscribers' mobile devices and the Internet. If carriers become mere data pipes—and some industry observers say they already are—carriers will have to compete largely on price, which will reduce their profits.

To minimize their dependence on Apple and Google, carriers are backing a range of new smartphone platforms. "There's an appetite in the industry to move away from the Android-iOS duopoly and really look at alternative operating systems," says Michael O'Hara, the chief marketing officer of the GSM Association (GSMA), which grew out of the coalition of European carriers that agreed to deploy GSM back in 1987, and is now the world's largest industry group of mobile-related companies. The new platforms come from either start-ups or established software providers that are expanding to smartphones. Most of them are licensing their software to multiple smartphone makers rather than making their own phones, which means they are more of a challenge to Android than iOS. Firefox OS is currently the most popular of these Android alternatives. Mozilla, the California company that makes the Firefox Web browser, commercially released Firefox OS in 2013. Some of the world's biggest carriers, including Spain-based Telefónica, are selling Firefox OS smartphones, and Sprint has expressed interest in eventually carrying them. Some large carriers are backing Tizen, as well. NTT DOCOMO, France's Orange, Korea's SK Telecom, and London-based Vodafone, the world's number-two carrier, are all members of the Tizen Association.

Windows Phone also has benefited from carriers' mistrust of Apple and Google. In 2013, Telefónica agreed to promote Windows Phone 8 smartphones across Europe and Latin America, stating that it wanted to "encourage the presence of additional mobile operating systems as an alternative to Android and iOS."[49]

SOFTWARE WARS

Smartphone software is more than just operating systems. Apps are key to the smartphone experience, and most operating system providers make at least some of their apps available on competing platforms. BlackBerry translated its BlackBerry Messenger (BBM) software for Android and iOS, and is doing so for Windows Phone. Google issues iOS variations of its main apps, such as Google Search, Gmail, Google+, and Google Drive, its service that stores and synchronizes files on remote servers in the cloud and enables access to them via the Internet.

Microsoft produces multiple mobile versions of Bing, its search engine; Office; and OneDrive, its online file storage service. Apple is the lone holdout. Apple reserves its software innovations for its own customers. If you want to video chat on FaceTime or talk to Siri, you must do so on an iPhone, iPad, an iPod, or a Mac.

Google is more open than Apple, but only to a certain extent. Its Maps and YouTube apps have provoked fierce skirmishes between smartphone companies. The clash between Apple and Google was the most public. In 2012, Apple dropped Google Maps and YouTube from the iPhone due to competitive concerns. Both apps had been mainstays on the iPhone home screen since 2007, but Apple debuted its own map and video software on the iPhone 5. Industry insiders interpreted the move as Apple's plan to create a "Google-free iPhone."[50] Not only was Google getting invaluable cachet from being the iPhone's default mapping and video-playing service, it was also collecting a huge amount of data from iPhone users' map and video searches. Google mined that information to improve its products and learn more about people's mobile habits.

Google was also using Maps as a weapon. Starting in 2009, Google steadily augmented the Android version of Google Maps with features such as voice-guided turn-by-turn navigation and fast-loading vector maps that could be saved offline. Google refused to license these add-ons to Apple, viewing them as a competitive advantage for Android. Apple's anger at being limited to an inferior version of Google Maps was one reason it removed the app from the iPhone.

Apple Maps was deeply flawed, and consumers lambasted Apple as soon as they realized its replacement technology was incomplete and unstable. Some Apple Maps directions suggested roundabout routes, Manhattan's bustling Lexington Avenue appeared in the neighboring borough of Brooklyn, and the Huey P. Long Bridge, which spans the Mississippi River, was located on dry land inside the city of New Orleans—to name a few examples. The maps also failed to locate major landmarks, such as Dulles International Airport, one of the Washington, D.C., area's main airports. While many of the errors were mildly annoying, some were dangerous, such as directions that instructed users to drive on train tracks and stranded people in the Australian desert.

The iPhone 5 still set a new opening-weekend sales record for Apple of more than 5 million units, but the public outcry eventually grew so loud that CEO Cook issued an apology. When Google released a downloadable, stand-alone Google Maps app about three months later, ravenous Apple users downloaded it more than 10 million times within two days. Google continues to distribute Google Maps and YouTube through the App Store as stand-alone apps.

In what has been dubbed "mapageddon," Apple and Google still compete fiercely over mapping technology, buying at least six mapping start-ups (total) in 2013 to bolster their technology arsenals. Both are motivated by the revenue potential of serving local ads to smartphone users who conduct searches via mapping apps. While the mobile maps–based ad market is small now—about 25 percent of all mobile ads, according to one estimate—it will certainly grow. As the

Washington Post has noted, "Just as the algorithm for finding information on the Web became the key to Google's search dominance, the precision and functionality of maps will be the key to mobile dominance."[51]

Google and Microsoft also fight over software. Their disputes have been less conspicuous but more hostile. Their most serious feud is over YouTube, since Google refuses to build a YouTube app for Windows Phones the way it does for iPhones. In 2011, Microsoft cited the situation as evidence of Google's anticompetitive behavior in a complaint to the European Commission, which at the time was investigating whether Google had violated European competition law in the search and advertising markets. Microsoft also raised the issue with the U.S. Federal Trade Commission. In 2013, when Microsoft created its own Windows Phone YouTube app, Google demanded Microsoft pull it, because it didn't comply with YouTube's terms of service. The skirmish attracted media attention, and the two companies appeared to compromise, publicly stating that they would work together to build a new Windows Phone YouTube app.

The brawl re-ignited several months later. In August 2013, Microsoft published a blog post entitled THE LIMITS OF GOOGLE'S OPENNESS, claiming that it had amended its YouTube app but that Google used software to block it upon release.[52] The Microsoft lawyer who wrote the post implied that Google was discriminating against Windows Phone because it uses Bing as its default instead of Google. In response, Google said it blocked the app because it continued to violate YouTube's terms of service. The Verge concluded, "Both companies are acting like children, and while the fight drags on, customers lose out."[53]

Microsoft appeared to capitulate in October 2013 by releasing an app that redirects users to YouTube's regular mobile website. As CNET wrote, the new app "jettisoned all of the extra features in prior versions . . . essentially negating the whole purpose of using a dedicated app in the first place."[54] In other words, the app wasn't really an app.

What's really behind the Google-Microsoft YouTube squabble? Google has said Windows Phone doesn't have enough users to warrant the resources Google would need to invest in developing apps. That argument is valid, but Microsoft is shouldering the bulk of the work and expense of the Windows Phone YouTube app, not Google. And Google does have at least one Windows Phone app: a version of its Google Search app. Perhaps Google simply wants to weaken Windows Phone by withholding one of its most popular apps.

DEVELOPER WARS

As platform providers vie to have the greatest number of app downloads, third party apps and the developers that make them have become so central to the smartphone business that companies spar over them, too. Apple is ahead at the moment, with more than 70 billion apps cumulatively downloaded since 2008, compared to more than 50 billion for Google. But due to its larger number of users, Google Play notched more than 15 percent more annual (not cumulative) downloads than the App Store in 2013, according to the mobile app analytics firm App Annie, and it is projected to pass the App Store in total app downloads soon.

Platform providers also compete over their total number of apps, which is one of the most closely watched and oft-cited metrics of a smartphone ecosystem's size and vitality. Google is currently winning this contest, stocking more than 1 million Android apps in Google Play as of July 2013. Apple hit the 1 million mark in October 2013. Microsoft and BlackBerry have far fewer: 245,000 and 140,000, respectively.

To narrow this gap, BlackBerry and Microsoft pay developers to build apps for their platforms. Vakulenko disapproves of the strategy, noting, "You can't buy developer love." Developers, he says, will take the money and churn out a so-so app. But BlackBerry and Microsoft have little choice; they need to give developers a reason to work with them. In the months leading up to its 2013 launch of BlackBerry 10 BlackBerry paid small developers $100 for every compatible app they wrote. The company ended up with 40,000 new apps. It also offered up to $9,000 for more sophisticated BlackBerry 10 apps. Microsoft reportedly pays well-known developers, such as the location-based social networking service Foursquare, $60,000 to $600,000 to make

Windows Phone apps.[55] Microsoft has also created apps for more than 90 popular companies by taking their mobile websites, repackaging them, and stocking them in its Windows Phone store. This initiative attracted controversy, since the companies, such as the ticket seller Ticketmaster and the hardware store Lowe's, didn't approve or even know about the apps.[56]

Apple and Google are in a different position. Rather than invest large sums of money in breaking through to developers, they just need to maintain their leadership as app ecosystems. Both offer developers marketing and promotional assistance to convince them to exclusively launch their apps in their stores. Apple also attracts developers by highlighting its status as the most profitable app store. All major app stores, including those run by Apple, Google, and Microsoft, take 30 percent of an app's sale price as a processing fee and give the remaining 70 percent to the developer. Apple regularly brags about the amount of money it has cumulatively paid out to its developers—$15 billion as of January 2014.

Google also publicizes Google Play's revenue growth, but in more abstract terms. For example, in April 2014, Google said it disbursed more than four times as much money to Android developers in 2013 than it had in 2012. It is difficult to estimate how much money that is, since Google doesn't specify how much its developers have earned in dollars. Google's reticence may be due to the fact that the amount is likely far lower than Apple's. The App Store makes more than twice the app revenue of Google Play, according to App Annie.[57]

Some of the most intense smartphone wars happen within app stores. Developers wage their own wars with each other. Smartphone ecosystems take care of one major developer worry: access to consumers. But stocking an app in a huge app store doesn't guarantee sales. Making a hit app is getting harder, due to a problem the industry calls "discoverability." In any app store a small group of apps, including the hottest games of the moment and the perennially popular, well-known services such Facebook and Pandora Radio, claim the lion's share of users' time and/or money. Developers outside this group must fight for scraps.

Developers are also having more trouble charging for their apps, due to the glut in the market, especially of free apps, which comprise more than 90 percent of all those downloaded. Developers give away their apps in hopes of making money through in-app purchases of virtual goods, game levels, and other extras or by convincing people to trade up to a paid, ad-free version. The trouble with these so-called freemium apps is that most consumers are content to stick with the free version.

App developers are a diverse crowd, from publicly traded companies such as Electronic Arts to precocious kids. The market analysis firm VisionMobile divides developers into eight groups: hobbyists are people who want to learn app development and have fun; explorers often work in the mobile or technology industry and are looking to make extra income on their own; hunters are professional developers who own or lead app development companies and seek to bring in as much money as possible with hit apps; guns for hire are professional developers aiming to attract commissioned or contract app work from other companies; gold seekers are start-ups focused on amassing a large audience for a free app as quickly as possible, in order to raise venture capital funding; product extenders are consumer goods companies that use apps to attract consumers to their main products; digital media publishers are magazines and news organizations that release their content through apps to lure readers; and enterprise ITs are large companies that create apps for internal employee usage. Of the eight groups, only hunters and guns for hire aggressively seek direct revenues from their apps.

Some hunter companies have minted hundreds of millions of dollars from their apps. The most notable are King, the Dublin-based producer of the candy icon-matching game Candy Crush Saga (though its 2014 initial public offering underperformed expectations); Finnish start-up Supercell, which makes the combat strategy game Clash of Clans and is valued at roughly $3 billion; and game publisher Rovio, creator of the blockbuster Angry Birds bird-and-pig–themed slingshot game. These success stories obscure the fact that 60 percent of the small, independent developers who want to make money off their apps earn less than $500 a month, according to VisionMobile.[58] Creating an app costs more than $27,000, on average, for iOS and more than $22,000 for Android,[59] due to the expense of paying developers and designers and purchasing tools, including computers, smartphones, and software. Struggling developers would need between four and five years just to recoup their costs.

Since consumers rely on short lists, for instance the top free apps and top paid apps in the App Store or Google Play, to filter good apps from bad, inclusion on these lists can increase downloads by many multiples. The App Store bases its "top apps" rankings primarily on an app's download volume and velocity (meaning the number of downloads in a given time period). Google Play weighs several factors, including app ratings (from user reviews) and an app's ability to attract and retain active users. Craig Palli, the chief strategy officer at Fiksu, a mobile app marketing firm that tracks app store trends, says the differences reflect the companies' backgrounds and strengths: Apple's as a seller of music, which traditionally gets ranked by popularity, and Google's as a search company that uses complex algorithms to rank results. The Windows Phone Store takes a hybrid approach, spotlighting highly rated apps alongside more conventional rankings.

The hurdle for new and obscure apps is getting onto the lists in the first place. Since 2009, more than 500 companies have released various app-related tools to assist bewildered developers, and this booming "SDK economy"[60] provides many ways to strategize an app's release, through ads, social media, and promotions.

To spur sales, a number of developers are buying so-called mobile app install ads that encourage consumers to download apps by providing direct links to app stores. Facebook pioneered these ads in its mobile app in 2012 and Yahoo, Twitter, and Google have created their own versions. Ads alone can't propel an app to the front of the App Store, though, so some developers break App Store rules by paying sketchy "app promotion" companies thousands of dollars to get their apps there. Silicon Valley's GTekna says it was the first app promoter to guarantee developers App Store placement. Starting in 2009, the company charged developers $7,000 to $15,000 to push an app into the App Store's top 25 over the course of two to three days. An app needed 50,000 to 60,000 downloads to make the list, says GTekna CEO Chang-Min Pak, so to marshal downloads the company resorted to bribery. It placed ads on websites that had large, active communities. People who clicked through the ads and downloaded apps were awarded points they could redeem for gift cards. In 2012 and 2013, Apple banned GTekna from the App Store and cracked down on similar apps.

A few upstart companies, including one called AppsBoost, have taken GTekna's place. AppsBoost currently charges developers $20,000 to drive an app into the App Store's top 25 free apps for a period of 48 hours, $12,000 to get an app into the top 50, and $8,000 for the top 100. An AppsBoost representative confirmed the company follows GTekna's old model of rewarding consumers with a piece of content or virtual currency for downloading a client's app.

Even shadier are the app promotion firms that employ "Internet water armies" of low-paid people in China and India to download apps. The name "water army" is a rough translation of a Chinese phrase and refers to these groups' ability to flood a specified website— or app—with high ratings, positive comments, votes, clicks, or downloads. Sometimes humans aren't even involved. There are firms in China that use automated software "bots" to download apps in massive numbers. When Flappy Bird, a free, ad-supported game that navigated a bird around obstacles, went viral in early 2014, people suspected bots had helped it notch more than 50 million downloads in just a few months.[61] Its developer has stated he said he doesn't "do promotion,"[62] but did not specifically deny the allegations and later pulled it from app stores.[63]

These "appayola" schemes are a niche business. VisionMobile, which tracks developer usage of third-party tools in annual surveys, says only 7 percent of developers reported they were utilizing any type of "cross promotion networks" to boost their apps in 2013[64] and the GTekna/AppsBoost brand of cross-promotion would be a smaller subset of that category. Such systems are ultimately unsustainable. "Once everyone [starts paying for downloads], the only way you can win is to put more and more money into it," says Vakulenko. "Eventually, the whole return-on-investment breaks down, and people start relying on it less and less."

Apple, naturally, despises appayola. Besides booting rogue developers from its iOS Developer Program, which essentially ends their careers as iPhone developers, Apple occasionally changes its App Store algorithms to thwart manipulation. Recent evidence suggests Apple's ranking now incorporates factors such as app rating, user retention, and user engagement as well as download volume and velocity,[65] though the weight it gives them is a matter of industry debate; Palli says Apple is "making tweaks," but has not instituted "core changes." The cat-and-mouse game between Apple and developers will continue.

FIGHTING FAKE PHONES

It's not just the app stores. Smartphone manufacturing and sales have a dark side. Counterfeiting and smuggling is pervasive in poorly regulated emerging economies. The fight to keep fake phones off the market is a front in a different type of smartphone war that pits opportunistic lawbreakers against smartphone makers and platform providers.

Smartphones are an ideal target for counterfeiters. All consumer electronics get knocked off, but smartphones are in a class of their own. Compared to other gadgets, they are in greater demand, command higher value, and are more portable. The global nature of the smartphone business, which spans hundreds of suppliers and subcontractors across multiple countries, also gives copycats ample opportunities to intercept phone designs and parts. As an April 2013 report about East Asian counterfeit goods from the United Nations Office on Drugs and Crime put it:

> Gone are the days when manufacturers could shutter their plants and guard trade secrets. Today, those who hold intellectual property rights are often situated half a world away from those who make their ideas come to life. . . . In effect, counterfeit goods are an untallied cost of the growth in offshore manufacturing.[66]

Mark Turnage is the CEO of OpSec Security, a company that develops and sells anticounterfeiting technology to governments and corporations, including smartphone makers. He says such protection is a growing and substantial market. Market research from IHS estimates 195 million "gray market" phones were produced and shipped in 2013. Since these phones are built without regulatory, durability, or safety testing, their quality is usually inferior. A 2011 Brazilian study of counterfeit phones found they took longer to connect to cellular networks and dropped more calls.[67] Manufacturers also sell knock-off phones illegally, evading import duties and sales and income taxes, and distribute them without warranties or customer service guarantees, leaving consumers vulnerable in case of problems after purchase. Market researcher Strategy Analytics estimates that smartphones account for 40 percent of gray market phones sold worldwide.

Counterfeits generally mimic a popular smartphone's exterior but contain substandard components. Counterfeiters may bribe a factory that does final assembly work on legitimate smartphones to run an extra "backdoor" shift and sell them the output. They may also obtain parts and designs from unethical component makers, corrupt employees, or by rooting through a component maker's trash. They can then manufacture the parts at a lower-cost facility or make them themselves. Whatever their approach, counterfeiters produce phones quickly and in large volumes to maximize their profits. "These are not mom-and-pop shops," says Turnage. "These are industrial operations that are manufacturing thousands of units. They are organized criminals with access to capital."

Neil Mawston, the head wireless industry analyst at Strategy Analytics, says China, Brazil, India, Indonesia, and the Philippines will be the world's largest markets for copycat smartphones through 2017. In some smaller countries, including Bangladesh and Sri Lanka, gray market phones are estimated to account for as much as half of all sales. Much of this production takes place in the city of Shenzhen, which is strategically located on the South China Sea near Hong Kong and Taiwan and has been a gadget manufacturing center for years. Gray market phones can be sold within China or exported—usually through Hong Kong to avoid paying value added tax (VAT), a consumption tax that many countries charge instead of sales tax.

Since the Internet makes it easy to source and sell counterfeits in any country, many knock-offs crop up online as well. Gray market phone sellers are active on consumer sites such as eBay and Craigslist. They also use business-to-business marketplace sites such as Alibaba that connect buyers and suppliers to sell knock-offs in large volumes to shady retailers. Brand protection companies, such as OpSec, track all of these listings, flagging smartphones that have unusual colors, features, or specifications. That could include an iPhone with dual Subscriber Identity Module (SIM) card slots, which would allow consumers to switch between cellular networks for lower rates or clearer reception or to keep a work number separate from a personal number. The dual SIM feature is popular in a number of countries, but Apple does not offer it on the iPhone.

Another tip-off is when a smartphone goes on sale online before its official launch. If the phone hasn't been released anywhere, a listing usually indicates a fake or possibly stolen phone. Phones that have only been released outside the country where they are being advertised are called "diverted" or "parallel imports." These are usually authentic phones sold at a markup (of up to 100 percent) to people who either don't want to wait for a phone's official release in their country or live in countries that get very limited supplies of the latest handsets—or never receive them at all. Diverters may illegally import the phone from afar. They may pay family, friends, or willing strangers, such as foreign exchange students, to obtain phones for them. Or they may travel to buy phones and smuggle them back. "Instead of selling a phone for $300 in the U.S., a diverter will openly offer it to anyone for $500 online," explains Turnage. "If you're a consumer sitting in, say, Chile, and you know this phone is not [officially] coming to your market for six more months, you might think, 'Why not buy it?' "

Demand for diverted iPhones is especially high in countries such as India and the United Arab Emirates immediately after new ones are unveiled. Apple did not ship the 5S to either of those countries until November 2013, and the staggered launch inadvertently created a lucrative, six-week window for dealers to smuggle them from the United States, Europe, and Hong Kong. iPhone diversion is so pervasive that Apple imposes restrictions on consumer sales. For years Apple required consumers to pay for iPhones with credit or debit cards, and to buy no more than two at a time during launch periods. In 2012, Apple amended the rule to ten per transaction.

Smartphone makers tend to handle counterfeiting issues on their own. To combat diversion and counterfeiting, some phone makers place tags, often printed with hard-to-copy holographic images, on newly assembled smartphones to ensure that they reach their intended destinations. If the smartphones are shipped internationally, customs administrators, with the help of detailed photo books provided by phone makers, can check the tags to distinguish authentic ones from knock-offs. A more sophisticated but controversial way to ensure smartphones aren't smuggled between regions is to place "region-lock" software directly on the devices. In late 2013, Samsung began programming some of its Galaxy Note II, Note 3, S3 and S4 phones to work only if users activated the phone on networks located in the same region in which it was purchased. Consumers complained that the feature presented unnecessary obstacles for frequent travelers and sometimes malfunctioned, preventing legitimate international usage of the phone, but Samsung defended the policy as "provid[ing] customers with the optimal mobile experience in each region including customer care services."[68]

Once smartphone makers have evidence of copying, they can initiate legal proceedings and contact law enforcement. To combat counterfeiters, smartphone makers can send cease-and-desist letters or work with law enforcement to stage raids and seize inventory. They can also alert eBay to counterfeit products; its Verified Rights Owner program lets brands log in to the site and flag suspicious listings, which eBay will then shut down.

Global enforcement efforts that require local police cooperation can be hit or miss. In the first quarter of 2014, Indian police conducted at least 13 raids in Hyderabad and other cities, which resulted in arrests and the seizure of more than $552,000 worth of fake Samsung phones and accessories. In recent years the central Chinese government has also tried to rein in counterfeiting operations through arrests and factory closures and in 2013 it coordinated with American customs officials on a crackdown that seized fake BlackBerry and

Samsung phones, among other products. But local governments are usually more permissive, due to corruption or economic concerns. "[Phone counterfeiting] is a net job creator," says Turnage. "So many people are employed in some part of it. Governments would need to replace those jobs with something else."

Some emerging trends will help smartphone makers fend off fake phones. Mawston, the Strategy Analytics analyst, expects gray phone manufacturing to peak in 2013 because legitimate phones are getting cheaper and more governments are clamping down on gray market sales. Countries ranging from Kenya to Pakistan to Uganda have either disconnected fake phones from their local networks, told carriers not to activate new counterfeit phones, ordered customs to reject phone shipments that haven't been preapproved by their telecom regulator, or plan to do so. Public safety is the primary motivation, since criminals tend to use fake, unregistered phones. Governments also want to curtail revenue losses from evaded VAT and import duties.

The global scope of the fake phone trade means reform will happen slowly. Countries that manage to delink fake phones often find users later reconnect them through software workarounds or resell them to neighboring countries. For now, fake phones make business even harder for companies embroiled in the smartphone wars.

4

Assembling a Smartphone

Smartphone users know little about how their phones get designed. The industry's cutthroat, litigious nature compels smartphone makers to guard their design processes zealously. Occasionally phone makers will share a few details about their design inspirations or briefly describe a new manufacturing process in their marketing materials, but for the most part consumers see only the finished product.

Some companies are more willing to talk than others. When asked to share how it designs smartphones, Sony offered a peek into its practices through interviews with three of its top designers: an art director, Rikke Gertsen Constein; a design manager, Tom Waldner; and chief art director Eiji Shintani. Constein, a Dane who formerly worked as a designer at Sony Ericsson, oversees Sony Mobile Communications' use of color and materials worldwide. Waldner, an American who has lived in Sweden and designed cellphones for decades, now manages Sony's three design studios—in Tokyo, Beijing, and the southern Swedish city of Lund. Shintani is Japanese and a prolific inventor who has 33 patents and patent applications on everything from phones to computers to radio receivers. He now directs the overall look and feel of a range of Sony products, including smartphones. The three of them help lead Sony's global smartphone initiatives and represent a range of backgrounds designers can have in a field that encompasses the globe and a multiplicity of skill sets.

In a conversation at Sony's gleaming high-rise Tokyo headquarters, Gertsen Constein, Waldner, and Shintani said Sony conceives sleek Android smartphones through a series of steps, including design research, individual brainstorming, and group feedback sessions. Each spring Sony's designers spearhead research they call "design intelligence" that informs their work for the subsequent year. Because

Sony designs its products for a future market, these forecasts look ahead two years so designers were working on their 2015 design intelligence in the spring of 2013. Cues can be found anywhere, but at Sony there are three main sources: sociological megatrends that are relatively constant but change slightly from year to year; consumer attitudes and behaviors; and design trends in other industries, such as furniture, which can be observed through events such as the Milan Furniture Fair. Sony has in-house researchers who do forecasts, but designers are expected to seek out trends themselves.

After several weeks of trend spotting, the designers distill their findings and produce a photo-studded PowerPoint presentation outlining what they consider to be the biggest global trends and the four to five main behavioral ones shaping the mobile computing devices industry. The document is shared with the entire team, which includes: industrial designers, who focus on the exteriors of phones; other designers who specialize in colors and materials; interaction designers, who create Sony-branded mobile applications; user interface and experience designers, who deal mostly with software; 3D modelers, who translate designs into software; 3D visualization designers, who produce still images and animations of phones; and graphic designers, who devise phone packaging.

The designers' research also produces a set of keywords that they are supposed to keep in mind while working. Past keywords have included "comfort" and "soft," and they represent some of the qualities people want in their smartphones. "Consumers would never say those words [to describe their ideal smartphone], but if you take all the research we do, we know this is what they're looking for," explains Waldner. "The keywords are not something we communicate externally [to the public] but something we ask designers to think about."

Once the intelligence research is complete, designers are briefed on the latest mobile components and materials, so they know what technologies are available to use. They also receive creative direction from the art director in charge of their particular discipline. They then jump off this "collective intellectual platform" into a six-month process called "design exploration."

Design exploration is when they start experimenting on their own to create Sony products. "Design intelligence says, this is what's happening in the world around us," explains Waldner. "Design exploration is each group asking, What does this mean for our product portfolio?" These products are primarily phones but can also be tablets or phone accessories, such as headsets and speakers.

Early phone designs can take any form. Shintani says some designers start with paper sketches. Others do 2D or 3D drafting on their computers or make shapes out of material such as cardboard. They share their ideas with their art directors and small teams, and after six weeks Sony holds a video conference for all its designers to present their work and get feedback. Presentations vary widely. Some designers may show up for the video conference with little more than a "mood board" of words and images that inspire them. Others will have an almost complete product design.

After the group conference, designers get four to six weeks to revise their ideas. About two and a half or three months into the design exploration phase, they choose which concepts to make into "gesture models," which are constructed of plastic, metal, and/or glass. These don't activate or run software; they are only intended to show the basic look of the phone. Once the models are ready, the designers hold more group presentations. Sony's marketers and technical teams look at the models and select which phones to commercially produce and offer their own input on changes. From then on, designers and engineers collaborate closely. Approved projects are given ready-to-launch deadlines and are assigned to a particular design studio to complete. Once the phone's design is finished, the device moves into manufacturing, and then sales.

Each smartphone maker has a unique design process. HTC CEO Peter Chou has described his company as "diehard about design."[1] HTC employs designers in its Taiwan headquarters and also has a Seattle-based studio, which specializes in user interface software, and a San Francisco industrial design firm called One & Co that it acquired in 2008. In the past, One & Co has designed snowboarding boots, watches, and furniture for companies including K2 Sports, Nike, and Timex. These other projects have informed the smartphone designs it creates for HTC.

Apple is notoriously private about its design methods, but details have leaked out over the years, often through former employees, and the company divulged some information in its 2012 patent trial with Samsung. Apple's process is highly centralized and iterative. According to longtime Apple designer Christopher Stringer, who testified during the Samsung trial, a small team of about 15 industrial designers determines the look of Apple's products. This group, which is based in Apple's Cupertino, California, headquarters, meets weekly, usually in a private kitchen connected to their design studio. Gathered around the kitchen table, the group opens their sketchbooks and trades ideas about all of the company's current projects. The best drawings are later imported into computer design programs and get made into 3D models directly in the studio, which boasts the latest rapid prototyping equipment so designers can quickly test their ideas. The results are shown to Apple's "highest levels of leadership," who ultimately decide which designs to pursue.[2]

Unsurprisingly, Samsung's design process is more similar to Sony's than to Apple's. Like Sony, Samsung operates design centers around the world, but on a larger scale, with seven centers, one each in London, Los Angeles, Milan, San Francisco, Seoul, Shanghai, and Tokyo. Samsung has said it employs 1,000 designers across its various gadget businesses, including its smartphones. The Seoul center is called the Corporate Design Center and serves as a central hub for conducting trend research and disseminating it to the rest of the company. In an interview published on one of Samsung's websites, a Corporate Design Center staff member described the process as follows:

> Each of [the other design centers] conducts its own research into what we call emotional values, elements that would appeal to consumers in two or three years' time. We then get together to review our separate studies and find commonalities. The selected emotional values—which would be a gauge of consumer taste two years from now—are then shared among Samsung designers and become the basis for a common design theme for all product sectors.[3]

To spur creative ideas, Samsung sends its designers on "experiential visits"[4] to exotic places. One designer translated his visit to a

Singapore hotel's rooftop infinity pool (built so that it appeared to extend to the horizon) into a visual "ripple effect" that made the Galaxy SIII's lock screen seem to undulate when touched.[5] A visit to Norwegian fjords influenced Samsung to issue the Galaxy S4 in a vivid light blue color called blue arctic.[6] Dennis Miloseski, the head of Samsung's North American design studio, recently told Engadget, "The design process [for Samsung phones] starts with a story."[7]

Sometimes that story stems from market research. Samsung gleans ideas for new phones from consumer polls and research reports: it was inspired to make the jumbo-sized, stylus-equipped Galaxy Note—sometimes called a phablet because its size is halfway between that of a phone and a tablet—after conducting global surveys. According to a 2013 *New York Times* article, "Samsung found that many respondents want a device that is good for writing, drawing and sharing notes. Asian-language speakers, in particular, found it easier to write characters on a device using a pen."[8]

Market research is a topic that divides smartphone designers. Steve Jobs famously said, "A lot of times, people don't know what they want until you show it to them," in a 1998 *Businessweek* interview.[9] Almost 15 years later Apple's head of marketing, Philip Schiller, reiterated this point while testifying during the 2012 Apple-Samsung patent trial. "We don't use any customer input in the new product process," he said. "We never go and ask the customer, 'What feature do you want in the next product?' It's not the customer's job to know. We accumulate that information ourselves."[10]

Sony conducts consumer research on smartphone designs for marketing purposes—essentially to ensure it puts the right products in front of the right people. Like other smartphone makers, it divides consumers into segments, such as the "mature tech" consumers who are tech-savvy and the "practical materialist" ones, who want basic functionality and good value. Sony does not convene focus groups to make design decisions. Doing that would "take the edge off our designs," says Waldner. "We would end up with a 'me too' product, a design everyone recognizes."

Yet consumer research has its place. Surveys helped Samsung accurately detect demand for very large phones. Though reporters mocked the first Galaxy Note's 5.3-inch screen as "awkward,"[11] "a brick,"[12] and even "a sure-fire recipe for ridicule,"[13] the phone was a surprise hit, selling more than 10 million units in less than a year. Market researchers at Canalys say one in three smartphones shipped today are phablets. Samsung, which was the first major smartphone player to supersize its phones, dominates this lucrative market.

Clearly there are multiple ways to design smartphones, and each process has its own merits. The world's best smartphones do have one characteristic in common: they were not easy to design. The Galaxy SIII went through 400 different design mock-ups. Apple will draft 50 takes of a single feature, such as a button, in search of perfection, according to Stringer. His colleagues, he says, are "a pretty maniacal group of people."[14]

DESIGN CHALLENGES

Smartphone designers need to be maniacal in some ways. Smartphones are among the world's most difficult gadgets to design, especially when engineering is taken into account. Jim Mielke, ABI Research's vice president of engineering, who specializes in smartphone engineering analysis, explains why: "Designing a smartphone takes so many disciplines: you need a good mechanical engineer to make the [outer casing] and the form factor [shape], which people are very sensitive to; you need an acoustic engineer to design how sound is picked up and how people will hear it; you need an engineer for not only the cellular communications, but also for the inside of the phone where you have GPS, Wi-Fi, Bluetooth, and NFC all working simultaneously. You need RF [radio frequency] engineering because there are so many frequency bands scattered throughout the world. You need someone who knows about smartphone 'brains' [processors] which are as powerful as most of the laptops people are using now. The whole thing has to be run by a tiny battery, so power management is another expertise you need. Smartphone imagery and video has become so elaborate, you need a good person for that to keep you competitive, too. And everything has to be cost-effective. Almost every single engineering discipline is required to make the thing."

There are other design challenges, too. Unlike a TV, smartphones must be aesthetically pleasing from all angles. As Shintani, the Sony art director, says, "Phones are not static; people look at them from every direction." Smartphones must also be ergonomically comfortable, because people hold them for long periods of time and cradle them against their heads. And since people are prone to dropping their phones, smartphones also need to be sturdier than TVs or laptops.

Smartphones create something of a design paradox. People want their displays to be as large as possible while still fitting into their pockets. To give people more space to watch videos, view photos, browse the Web, and play games, manufacturers have progressively made phones longer and slightly wider. The other way to make a display larger is to reduce the width of the bezel, the black rim surrounding the display. That enlarges the size of the touchscreen without affecting the phone's overall dimensions. Shrinking the bezel is complicated, and smartphone makers do it only on their most expensive phones. High-end displays are so thin—"almost like a human hair," according to Mielke—that if they are not secured, they can twist or crack and break upon impact.

Even though screens are getting larger, people still want their phones to be as slim, light, and as strong as possible. To satisfy these demands, smartphone designers and engineers have compressed phones to the point that some are thinner than 6 millimeters. The trick is fitting all the necessary components into a narrow casing, since consumers also want their smartphones to be fast and have a long battery life. To achieve this, companies have redesigned some parts. Samsung made its 2011 Galaxy SII more than a millimeter thinner than the original Galaxy S by reducing the number of layers in the phone's display and streamlining the camera. To get the iPhone 5 18 percent thinner and 20 percent lighter than the iPhone 4S, Apple took similar steps: shrinking the phone's SIM card and camera module, among other parts, and baking the phone's touch circuitry directly into the phone's display. More power-efficient displays are also helping smartphone makers prolong battery life without resorting to larger-capacity (and bulkier) batteries. Mielke says Motorola's flagship phone, the Moto X, has a particularly economical display that draws just 92 milliamps of electric current at its brightest. A few years ago smartphone displays drew 700 milliamps, or about eight times as much current under similar conditions.

To improve smartphone durability without adding heft, designers have been experimenting with metals, such as aluminum and titanium, that are sturdy as well as lightweight. These designs require special attention, since metal can interfere with the transmittal of radio frequency signals—a problem that affected the iPhone 4, sparking a controversy in 2010 dubbed "Antennagate." According to Walter Isaacson, Antennagate was a case of design versus engineering. Apple design director Jony Ive gave the iPhone 4 a steel rim, thinking it would fulfill three purposes: "be the [phone's] structural support, look really sleek, and serve as part of the phone's antenna." [15] But when the iPhone 4 was held a certain way, the user's hand covered a small gap in the rim, causing signal loss and, sometimes, a call to drop. Apple had to issue free phone cases to mollify irate iPhone owners, and it later paid another group of iPhone users $15 each to settle a class action lawsuit that alleged the company had misrepresented the phone's antenna and reception quality.

After Antennagate, engineers developed technology that can sense where the user's hand is located and send and receive signals from different areas of a phone based on that information. Phone makers also insert small strips of glass or plastic into their metal smartphones to enable signal transmissions. "Metal is a good smartphone material," says Mielke. "You just need someone who knows how to make it work." Designers and engineers are known to disagree often, but smartphones are such complex devices that the two groups must collaborate to ensure their phones will work as intended.

As smartphones get thinner and flatter, designers literally have less area to be creative. Cellphones used to come in a variety of shapes: the straight, simple candy bars Nokia made famous; the Motorola RAZR-like flip style that folded in half and flipped open; and Sidekick-like sliders, in which the screen slid up to reveal a keyboard. Smartphones increasingly look the same: thin slabs with large touchscreens on one side and a camera lens on the other. Designers must devise ways to distinguish their phones without adding thickness or weight or deviating from the core specs consumers demand. Some designers have turned to bright colors, custom-designed casings, and textured plastics to differentiate their smartphones. Plastering a jumbo logo onto the phone is not a viable solution. "We are proud of our brand, but if we make [the logo] big, it takes away from the elegance of the product," says Gertsen Constein, the Sony art director.

Carriers are often a factor in smartphone design. Since carriers and retailers sell the majority of smartphones, they usually have influence over the design process (except at Apple). Sony shows carriers its design concepts at an early stage to get feedback. Waldner says this "early engagement process" serves as a reality check for designers. "If a [carrier] says, 'This is terrible,' then we would do something to fix it," he adds. "We have killed some projects in the past; not many, but some."

Smartphones that aren't tailored for a particular carrier need to account for regional preferences yet still appeal to as broad an audience as possible. Waldner says European consumers are usually more interested in the ways a phone will fit into and improve their lives, while Asian consumers tend to pay more attention to phone specs. Creating phones that attract both audiences is complicated but necessary, since smartphone makers have little control over the way carriers market their devices once they buy them. "[Different] operators have different socioeconomic situations," points out Waldner. "A relatively inexpensive phone may be sold to ambitious [professional workers] in India, but in Europe it might be positioned to young people."

The industry's rapid pace is another challenge for designers. Sony works with a range of carriers and many purchase phones during different months. To coordinate with various carrier schedules, Sony designs new batches of smartphones twice a year. Sony's design process might seem leisurely from the outside, but the company's designers liken it to a sprint, because so many factors and teams (design, engineering, marketing) need to come together before a phone can be completed and sold.

Marko Ahtisaari, Nokia's former head of design, lamented the industry's whirlwind pace in a 2012 *Guardian* interview. Asked about Nokia's upcoming smartphone designs, he said, "This is a funny industry. As soon as you see something beautiful the question is 'what's next then?' We have something beautiful on the table; let's just enjoy it." [16]

TEARDOWNS

Design is only one element of a smartphone's appeal. Of course smartphones also need to have the right components. Smartphone makers and component suppliers have a synergistic relationship. Each business relies on the other, and their relationship yields innovations that get incorporated into phones. In the early days of smartphones, component suppliers often concocted new technologies, such as sensors, and pitched them to smartphone makers, which sometimes didn't know what do with them, says Mielke. These days competition is so fierce that smartphone makers commission component suppliers to develop brand-new technologies dreamed up by their designers, engineers, and marketers. "Smartphone makers are asking, 'What else can we make this device do?' " says Mielke. "They go to component manufacturers and say, 'Here's what we need, can you make it?' "

These discussions, like all components negotiations, take place in secret. Smartphone makers rarely reveal their components' suppliers and never disclose the cost of their phone parts. In an industry full of price wars and copycats, component data is viewed as extremely valuable and confidential. "[Phone makers] want to make it as hard as possible for competitors and customers to figure out what's inside their phones," says Greg Bridgett, a former product manager at the technology analysis firm UBM TechInsights, which researches smartphone manufacturing techniques. "That allows them to come in with premium pricing." Mielke says that smartphone makers are also concerned about competitors imitating their designs and engineering and deliberately disrupting their access to components. "The more secrecy you have, the safer you are [in the smartphone industry]," says Mielke. "It's about protecting yourself."

To protect themselves from warranty and injury claims, smartphone makers also make it difficult for consumers to explore the insides of their phones. Some devices, including the iPhone, can't be opened without special screwdrivers, because they have screws in unusual shapes. iPhone screws have five points, like a star, instead of the common single slot or two crossed slots, like a Phillips head screw. Even smartphones that have removable back covers usually only let users access their batteries and SIM cards.

Fortunately, for the curious, there are technical analysis and reverse-engineering firms that dissect smartphones and other devices and analyze what they find. These analyses, called teardowns, catalog the smartphone components they uncover and list who manufactured what. Professional teardown reports often cost a few thousand dollars and include component photos, descriptions, comparison tables, and access to an online database of teardown data. But reports can vary greatly in detail, content, and price. An in-depth reverse-engineering report that evaluates the circuits inside every chip in a particular smartphone, takes 18 months to write, and is provided on an exclusive basis to a single customer could cost as much as $250,000, says Tom Cox, the former marketing director at Chipworks, an Ottawa-based reverse-engineering firm.

Teardown firms sell their data to an array of customers. Smartphone companies read teardowns for technical information because they want insight into their rivals' businesses: Are they using new chips? If so, who made them and how expensive are they? Carriers use teardowns to gather data about smartphone costs, which help them negotiate purchasing contracts with phone makers. Financial analysts buy teardowns to gauge company margins and profits and make investing recommendations. Patent holders and lawyers look at teardowns and reverse-engineering data for proof that companies are infringing their intellectual property and should be paying royalties. Teardown firms also publish abbreviated versions of their most popular reports online or share them with reporters, which is usually how teardown data becomes public.

In pursuit of information, teardown firms sometimes act like detectives. Most smartphone parts are unlabeled, and the ones that are labeled are usually marked with a string of letters and numbers rather than easily identifiable names. Researchers have to run those letters and numbers through search engines and special databases to figure out who made the parts and sometimes how the parts function within the phone. Unmarked chips are particularly confounding because they look alike from the outside. Identifying them requires specialists with engineering and advanced materials science training. Kyle Wiens, the CEO of iFixit, a California company that encourages people to fix their own gadgets by publishing repair guides, selling repair tools and replacement parts, and performing teardowns, calls this work "digital forensics."

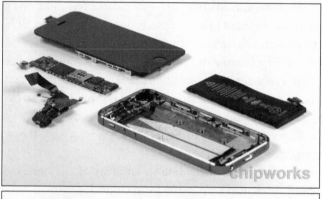

[Top] iPhone 5S teardown, part 1 (*Chipworks*)
[Bottom] iPhone 5S teardown, part 2 (*Chipworks*)

Teardown firms can tell where a chip was made based on how it was made. Companies such as Chipworks will submerge chips in acid to determine their provenance. The acid dissolves its plastic outer packaging so engineers can inspect the shiny, metallic die inside. A die will typically bear some printing from the chip's manufacturing process or a logo or image, and if the markings on the die can be deciphered, they will often indicate the maker and sometimes its part number. Teardown firms can also deduce a chip's origins by comparing its layout design to other chips and gauging the sophistication of its manufacturing process. Only a handful of companies in the world have the infrastructure capable of producing the tiny, extremely power-efficient chips used in very high-end smartphones.

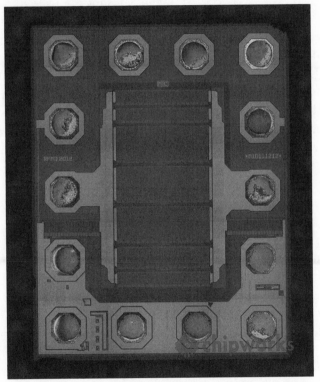

Interior die photo of the iPhone 5S antenna tuning/switch chip (*Chipworks*)

Smartphone companies may not like teardowns, but they can't crack down on them, because reverse engineering is legal in the United States, Europe, Japan, and other regions. The Semiconductor Chip Protection Act, which the U.S. Congress passed in 1984 to protect chip designs that have been registered with the U.S. Copyright Office, allows reverse engineering "for the purpose of teaching, analyzing, or evaluating the concepts or techniques embodied in . . . [a chip's] circuitry." [17] Under that and similar laws around the world, Chipworks's Tom Cox says, "reverse engineering is legal at all levels of analysis; it is copying a device, building a knock-off, and infringing on others' patents that would be an issue."

Teardowns reveal some interesting facts about smartphone costs. Teardown experts say HTC, LG, and Samsung are more likely than Apple to snap up pricy new components for their flagship smartphones. Market research firm IHS pegs the iPhone 5S's hardware cost at $191 and the Galaxy S4's at $236. The Galaxy S4's bill of materials is high because it incorporated some unusual chips, including sensors that measure atmospheric pressure, humidity, and temperature. "Usually you see one to two new things inside a new phone," says Rob Williamson, a former Chipworks marketing manager. "The Galaxy S4 had five." The Galaxy S5's bill of materials cost was even higher: $252, partly because Samsung added two more sensors, for detecting users' fingerprints and heart rates.

FOLLOWING THE SMARTPHONE MONEY

The smartphone market is global, but the hardware comes primarily from three regions: East Asia, the United States, and western Europe. Mielke estimates the average smartphone contains 800 to 1,200 components. Nomura Equity Research's annual "Smartphone Guide" report, which lists the world's leading smartphone component suppliers as a resource for investors, provides a filter for examining this sprawling market. In its 2013 report, Nomura identifies 64 key suppliers for 24 types of smartphone components. Samsung is cited as the single largest supplier, both by breadth and by market share. Counting its subsidiaries, Samsung makes 12 different types of components that are covered in the guide, from displays to batteries, and holds 45 to 50 percent of the market for some of them. For instance, it is the world's largest supplier of smartphone memory chips.

In terms of countries and regions, most smartphone component suppliers are based in the United States or East Asia; western Europe plays a smaller role. The Nomura guide mentions eight European companies, primarily based in Austria, Britain, France, Germany, the Netherlands, and Switzerland. "It's a very complicated supply chain," says Chipworks's Williamson. "It's a global business with money and people crossing everywhere."

Understanding where the money flows in smartphone manufacturing requires looking at these companies, or at least the ones that produce higher-priced components. The parts included in Nomura's guide cost as little as $0.40 and as much as $40. A smartphone's most expensive component is often its display, and a high-end one comprises $30 to $40 of a smartphone's overall cost. These boast sophisticated technologies, such as IPS (in-plane switching) and AMOLED (active-matrix organic light-emitting diode), that show brighter, sharper images and more true-to-life colors. Display technology progresses so rapidly, and image quality has become so integral to the smartphone experience, that even low-end and midrange smartphones need high-definition displays to be competitive. "Today, if [you're a smartphone maker and] your smartphone doesn't have 1080p or 720p, meaning the same number of pixels as a good, large-screen TV, you're nowhere to be found," says Mielke. Up next: so-called 4K displays that have four times as many pixels as a 1080p TV, and thus four times more resolution.

Memory chips are frequently the next largest smartphone expense, particularly in high-end phones that need a lot of memory to run smoothly. Nomura estimates that memory chips add another $30 to $36 to a phone's cost. That figure represents a medium amount of memory—a combination of 32 GB of flash memory at $20 to $24 and 2 GB of dynamic random-access memory, which is often just called RAM, at $10 to $12. (The iPhone 5S comes with 1 GB of RAM and either 16, 32, or 64 GB of flash memory, and the Galaxy S5 comes with 2 GB of RAM and either 16 or 32 GB of flash memory.) Smartphones use flash memory for storage and RAM for processing. Memory helps smartphones load apps, games, and websites faster and to juggle multiple tasks with less time lag.

Nomura says a smartphone's applications and baseband processors are the next priciest parts of a smartphone. Both types are essential to its operations and are complicated to build. The applications processor—often called the phone's brain—is the smartphone's central processing unit (CPU) and functions much like a computer CPU, controlling all of the phone's major operations. A smartphone's baseband processor handles the radio functions that allow the phone to connect to cellular networks. Nomura estimates each of those processors contributes an additional $16 to $18 to the phone's bill of materials. But advanced applications processors often cost between $30 and $40, so some research firms consider them to be the number-one or -two smartphone component cost. "In a superphone, memory will be up there in cost, but not anywhere near the processor," says Bridgett.

Most of the money in high-end components goes to the United States and East Asia. A San Diego–based company called Qualcomm is the world's largest maker of smartphone applications processors. Market researchers at Strategy Analytics says Qualcomm scoops up 54 percent of the revenues in this lucrative market. Qualcomm isn't a household name, but it has been a leader in cellular technology since the 1990s, when it invented the algorithms behind CDMA. Mielke says Qualcomm dominates processor sales because its "continual, aggressive investment in and advancement of technology" gives its chips superior capabilities. Qualcomm spent close to $5 billion on R&D in its fiscal 2013 year, which was less than Samsung, Microsoft, or Google but slightly more than Apple. Every Windows Phone runs on a Qualcomm Snapdragon processor. So do smartphones such as the HTC One (M8), LG G2, Moto X, and the Sony Xperia Z2. Qualcomm is also the largest seller of cellular-baseband processors and commands a 64 percent revenue share of that market.

In other areas of smartphone componentry, Asian companies dominate. The main manufacturers of smartphone displays are Korean, Japanese, or Taiwanese. Korean and Japanese companies are the largest makers of smartphone memory chips. According to Rich Howard, a research scientist at Rutgers University's Wireless Information Network Laboratory, this is because producing chips and displays involves completing a complex series of manufacturing steps with absolute precision in high volume; government support for Japanese and Korean manufacturing companies that bolsters them with R&D funding and worker training partly accounts for their abilities to thrive.

European companies lead in a few smartphone component markets. Scotland's Wolfson Microelectronics sells a third of all smartphone audio chips, according to Nomura. Dutch chip maker NXP Semiconductors provides about half of the near field communication (NFC) chips in smartphones. (NFC enables contact-less mobile payments between phones and payment terminals. When you swipe your phone to pay for a product in a store, you're using NFC.) Geneva-based STMicroelectronics supplies about half of the gyroscope and accelerometer sensors that help smartphones detect speed and orientation for use in games, navigation, and other applications. But these chips and sensors cost just a few dollars each, or even less. In general, American and Asian companies reap more revenue from smartphone components than European companies.

INSIDE THE IPHONE

Suppliers don't profit equally from smartphones. Figuring out how much each company or country really earns from a particular smartphone requires deeper analysis than just looking at teardown data. In 2011, three researchers, one each from Syracuse University, the University of California, Berkeley, and the University of California, Irvine, conducted just such an analysis of the iPhone. To determine which countries prospered the most from the iPhone's popularity, they calculated what each supplier earned from selling its components. They then deducted the expense of producing those components to get the gross profit—the amount a company earns from making and selling a product that does not include its overhead and sales expenses.

Using this methodology the researchers found that Apple earned $321 in gross profit from each iPhone 4, which is equivalent to 58 percent of the phone's retail price. In comparison, Apple's Korean suppliers made $26 in gross profit from each phone (a 5 percent share of the sale price). Suppliers in other countries made even less. Suppliers from the United States earned $13 (2 percent), Europe $6 (1 percent), and Japan and Taiwan, $3 each (less than 1 percent). In addition, un-identified suppliers earned $29 in gross profit (5 percent).

In other words, all of Apple's suppliers earn money from the iPhone, but the amount fluctuates based on the type of part they supply, and none of them earns anywhere near as much as Apple. The study concluded, "While other companies are thrilled to be part of the supply chain for these highly successful products, their benefit in dollar terms pales in comparison to Apple's." [18]

Apple profits from the iPhone because it maintains control of the highest-paid parts of its creation. That includes the design, software development, product management, and marketing, all of which it does in-house. Making and selling components is less profitable work, and that's what Apple outsources.

A couple of caveats should be noted regarding this study. It placed costs for the materials used to make the components in an "other" category, because the researchers didn't know where (meaning, in what country) the materials were produced. Those materials make up $120, or 22 percent, of an iPhone 4's retail price. Jason Dedrick, who teaches information studies at Syracuse University and co-wrote the iPhone study, says these transactions are less important because they are less profitable. "For the most part, these are commodity goods, so the margins are probably not that high," he explains.

The study's numbers are also specific to the iPhone 4, which was the newest iPhone available at the time. Later iPhones have different costs and components and some different suppliers. Nevertheless, Dedrick says the study's overall conclusion remains the same: the iPhone is a big moneymaker, mostly for Apple.

SUPPLY CHAINS

Apple's grip over its suppliers has propelled the company's financial success. Carefully managing its supply chain gives Apple more control over its component supply and pricing, which lets Apple keep its iPhone production expenses low. Supply chain management also helps Apple speed iPhones to market with leading technologies, enabling it to sell at high prices and more units overall. Apple profits both ways. As John Gruber, the influential Apple commentator, put it, "However superior Apple's design is, it's their business and operations strength . . . that is furthest ahead of their competition, and the more sustainable advantage." [19]

Apple enjoys several operational advantages over its competitors. Stan Aronow, who analyzes technology companies' supply chain practices at the research firm Gartner, says Apple wields its prestige to convince suppliers to produce entirely new components for iPhones. What's more, Apple gets the components customized to its exact specifications in perhaps half the time such a process would normally take. To meet Apple's deadlines suppliers will assign their most talented engineers to the project and ask them to work nights and weekends, says Aronow. Suppliers make these sacrifices because iPhone contracts are huge (due to large volumes, not high margins). Collaborating with Apple is also prestigious and can bring in other business. "There's a halo effect to [being an Apple supplier]," says Aronow. "Companies can say, 'Hey, we did this for Apple, we can do this for you.' "

The drawback of being an Apple supplier is overdependence. Apple is such a large, demanding customer that some suppliers derive 50 percent or more of their revenues from the company and its i-devices. If iPhone sales slow, and Apple cuts its orders or fully terminates its contract, suppliers' stock prices can plummet, and companies can even go bankrupt. Shicoh Co., a midsize Japanese company that made small motors for iPhones, filed for bankruptcy protection in 2012 after it overestimated the size of its next Apple order and overspent to increase production.

Such slumps are less of a concern for a select few suppliers that Apple gives special treatment so it can lock in access to important iPhone components. Apple does this in several ways: loaning these suppliers billions of dollars, so they can expand their facilities; purchasing equipment for them; and paying them ahead of time for the future delivery of components. In 2013, Apple CFO Peter Oppenheimer told analysts, "We are buying equipment that we will own, that we will put in partners' facilities. Our primary motivation there is for supply, but we get other benefits as well." [20] One of those other benefits is likely exclusivity—not only exclusive use of the equipment it buys but also exclusive access to technological processes and materials its suppliers develop. Asymco analyst Horace Dediu, who has analyzed these investments in depth, notes, "Apple has a clever way to 'own factories' without actually owning them." [21] He estimates that Apple has invested close to $30 billion in machinery since launching the iPhone in 2007.

Apple's most recent such deal involves GT Advanced Technologies, a U.S.-based supplier of synthetic sapphire glass. Apple already uses this tough, scratch-resistant glass in its iPhone camera lenses and the iPhone 5S's Touch ID fingerprint sensor and it may apply the glass to strengthen future iPhone screens, as well. To ensure adequate supply of this unusual material, Apple is prepaying GT $578 million to buy furnaces for producing the synthetic glass. GT is repaying Apple that money over five years, starting in 2015, and has agreed to make sapphire exclusively for Apple for a certain amount of time. [22] Aronow says Apple invests wherever it thinks it needs to in order to be first to market with leading technologies. "If that means they need to invest in proprietary equipment at their suppliers, then so be it," he says.

Samsung is the only smartphone maker that controls its suppliers to the extent Apple does. Gartner has praised Samsung's ability to adjust its manufacturing and inventory on a weekly basis "despite its massive size and product diversification." [23] Samsung is agile because it is its own supplier. It manufactures the Galaxy S4s and S5s displays and memory chips and produces most of its applications and baseband processors. The U.S. version of those phones uses a Qualcomm applications processor and baseband processor. The international version of the phones uses a Samsung applications processor and a baseband processor from computer chip giant Intel. In addition, some Galaxy S4 phones use Toshiba memory chips. Nevertheless, analysts have estimated Samsung pockets more than 80 percent of the component profits generated by Galaxy S4 sales. [24]

Samsung's close contact with retailers and distributors also helps it react quickly to market demand. Aronow says it has set up special software systems at its partners' locations so that it knows when its smartphones are running low in a particular store or zip code and when they are piling up at a carrier or retailer. To get sales moving again, Samsung will launch a targeted promotion, such as a temporary price cut on a certain smartphone. "Samsung runs a weekly process of matching supply and demand," says Aronow. "It's a lot more sophisticated than what we see from other smartphone supply chains . . . that are more one-size-fits-all."

Smaller smartphone makers have much less control over their suppliers. These companies may run out of components and lose sales to Apple and Samsung. When HTC planned to launch its One phone in March 2013, insufficient supply of camera parts held up shipments through May. During the delay, Samsung released its Galaxy S4, overshadowing HTC, though the One was widely regarded as a better smartphone. Ben Wood, an analyst at the market researcher CCS Insight, summed up the situation well: "Samsung do[es] this blitzkrieg marketing. Poor little HTC lost their window of opportunity."[25] The botched One launch led HTC to report a record-low quarterly profit in April 2013 and, later that year, its first-ever quarterly loss (of more than $100 million).

Even when component quantity is sufficient, smaller players lose out on price. In manufacturing, large orders earn discounts, and costs decline as production increases; Apple and Samsung enjoy economies of scale. Apple, in particular, is known to be a hard bargainer with suppliers. HTC and other smartphone companies lack the clout and volumes necessary to negotiate low component prices. Smaller companies also pay more to ship their phones to carriers and retailers. In the smartphone industry, speed and scale separate winners from losers.

CONFLICT MINERALS

While smartphone design often starts with a sketch, and engineering often starts with the selection or commission of components, production could be said to start with the mining of minerals that go into the components. Much of this mining takes place in Africa, and it can carry a human cost.

Activists say cellphones and smartphones are fueling the trade of four conflict minerals that get processed into the metals tantalum, tin, tungsten, and gold, sometimes collectively called "3TG." The Democratic Republic of Congo exports all four of them, and is one of the world's leading suppliers of tantalum. The DRC is not the only country with 3TG reserves, but minerals there are financing the violence, brutality, and chaos that give conflict minerals their name. Armed Congolese groups have hijacked control of the mining and sale of these lucrative minerals, which is estimated to amount to hundreds of millions of dollars per year, and they use the profits to illegally buy guns to maintain their power. More than 20 rebel groups are believed to operate in the DRC, destabilizing the region and committing serious human rights abuses, including rape and other forms of extreme violence.

Marcus Bleasdale, a photojournalist who has been covering the conflict minerals issue since 1999 for publications including *National Geographic*, says that while conditions in much of the DRC have improved dramatically since he first started visiting the country, terrible strife remains in the eastern part of the country. Worst off are a region called Ituri and the North Kivu and South Kivu provinces, which have rich mineral deposits and are located close to countries, such as Rwanda and Uganda, that allow smuggling through their porous borders. As Bleasdale says, the DRC's plight is not just about natural resources; it's also the result of rebel control of trade routes and ethnic conflict between Rwanda and the DRC.

Bleasdale, who regularly visits to report and take photos, describes parts of the country as "desperately sad." More than two million have fled their homes to escape rebel fighting and encroachment. These refugees crowd into makeshift camps, where they live in thatched huts covered with plastic sheeting and receive rations of rice and beans. In rebel-controlled areas, hospitals have been looted of everything from medicine to mattresses. "People will sit in a hospital just waiting for someone to show up with some medicine so they can be treated," says Bleasdale.

He has also seen children as young as seven forced to work in rebel-controlled mines after warlords seize their villages. The children get paid a nominal amount of money but "are essentially working because of fear." In these mines labor is manual, rather than mechanized, and usually takes place in large, open pits. When tunneling is necessary, children often get pressed into service, because they are small. "A seven-year-old can shovel dirt at the base of a cave," says Bleasdale.

"Children aged eleven or twelve can be sent down into tunnels to chip away at a rock face or to carry rocks from the center of a mine to the outside."

Tin mine in North Kivu, Democratic Republic of the Congo (*Sasha Lezhnev/ Enough Project*)

These extracted minerals are used in a number of consumer electronics, such as laptops, iPods, and digital cameras. Cellphones and smartphones use all four of the 3TGs. Tantalum is used to build capacitors, which store electrical energy. Phone makers use tin, along with lead, to solder parts onto circuit boards. Gold is used to connect components on circuit boards, because it's good at conducting electricity. Tungsten, which is very dense, is used to help make phones vibrate.

Since the amount of conflict minerals in a device usually corresponds to its size, a single smartphone doesn't use a lot of these minerals. But because cellphones and smartphones are produced in massive quantities they are believed to be one of the top destinations for conflict minerals. Furthermore, smartphones contain more conflict minerals than basic phones, because they have larger circuit boards and more components.

Conflict minerals take a circuitous route from the DRC to our smartphones. First minerals leave the mines as ore. Armed groups smuggle the ore east into other African countries, and from there it is shipped to smelting companies in China, India, Indonesia, Malaysia, and Thailand. They melt the tantalum, tin, and tungsten ore to extract metals and refine the gold ore to remove impurities. The results are then sent to manufacturing companies, usually in Asia. These manufacturers mix the metals with other metals to create materials such as tin solder. Circuit board manufacturers buy those metals for their boards, and in the final step, phone makers or their contract manufacturers place those boards inside phones.

Activists trying to stop the atrocities plaguing the DRC want technology companies to patronize conflict-free smelters and component suppliers that pledge not to buy conflict minerals. "Natural resources are the cornerstone of this region's local economy," says Bleasdale. "We have to find smart solutions, so the Congolese can benefit from their natural resources, as opposed to blocking any benefit from their sale and profit."

Conflict-free sourcing requires tracking minerals from the mine, certifying the smelters that melt and mix the minerals, and identifying which suppliers buy metals from which smelters. This work is already under way. The DRC government and various nongovernmental organizations (NGO) have established "bag and tag" systems that affix plastic bar codes to bags of minerals before they are traded and transported to smelters. The U.S.-based Electronic Industry Citizenship Coalition (EICC) and the Europe-based Global e-Sustainability Initiative (GeSI)—two global coalitions of electronics companies and related organizations that deal with corporate responsibility supply chain issues—have set up a voluntary smelter certification program. So far 84 smelters have been certified as conflict-free, out of an estimated 300-plus in operation worldwide.

These efforts have reduced the profits rebel groups make from selling untraceable tantalum, tin, and tungsten by approximately 65 percent since 2010, according to the Enough Project, a Washington, D.C.–based nonprofit dedicated to fighting crimes against humanity, including conflict minerals. Because more and more companies want to buy only clean minerals, mines free of armed control can fetch $9 for a kilogram (2.2 pounds) of cassiterite (tin ore), whereas mines controlled by armed groups are only able to get $1.50 per kilogram, Bleasdale says. Enough says 66 percent of Eastern Congo's 3T mines are now demilitarized.

Apple has become the most proactive smartphone maker on these issues, according to Sasha Lezhnev, the Enough Project's senior policy analyst. Apple used to be a laggard on conflict minerals—so much so that a coalition of activists, including the Enough Project, staged a protest at the opening of the company's first Washington, D.C., store in 2010. Apple gained credibility with activists in 2013 when it signed on to the Public-Private Alliance for Responsible Minerals Trade (PPA)—a group of governments, NGOs, and companies that is funding conflict-free initiatives. Apple also joined the Conflict-Free Tin Initiative (CFTI), which is comprised of companies building a conflict-free tin-sourcing system in eastern DRC, and in February 2014 Apple declared its tantalum smelters clean (based on third-party audits) and said it was "pushing [its] suppliers of tin, tungsten, and gold just as hard to [also] use verified sources." [26]

Other smartphone makers have yet to verify their products as free of a particular conflict mineral, but BlackBerry and Nokia have also been active in helping to set up certification and tracing systems. Both companies are part of the PPA, the CFTI, and, along with Motorola, participate in the Solutions for Hope project, a corporate alliance that is devising ways to source conflict-free tantalum from the DRC.

Conflict minerals have been in the news since the 1990s, but most American companies ignored the issue until 2010. That year the Obama administration passed the Dodd-Frank Wall Street Reform and Consumer Protection Act that proposed regulations for corporate use of conflict minerals, among many other financial reforms. When the U.S. Securities and Exchange Commission (SEC) adopted its final rule on conflict minerals in 2012, it made them a core business concern for many companies. Starting in 2014, public U.S. companies must tell the SEC whether conflict minerals are "necessary to the functionality or production" of products they manufacture or have hired other companies to manufacture for them. [27]

Smartphone companies that aren't publicly listed in the United States don't have to file these reports, but many foreign tech companies are taking action anyway to avoid harsh criticism from activists, nonprofits, and socially conscious consumers. Lezhnev says tech companies in general are worried that any connection to conflict minerals will hurt their brands—similar to the damage that allegations of sweatshop labor or child labor can wreak. Such anxieties "make sense," Lezhnev adds. "[Technology] is a highly competitive industry, and these companies have to deal directly with consumers."

No matter how zealous they get, smartphone and electronics makers won't be able to halt the conflict minerals cycle on their own. Illegal trade of the 3Ts is down, but conflict gold has been harder to rein in, because it is hugely profitable, even in small quantities. The Enough Project estimates $500 million of gold dust and nuggets are smuggled out of the DRC each year, often hand carried on commercial planes in quantities of less than 100 pounds to countries such as the United Arab Emirates for purchase and resale. Jewelry companies are the biggest buyers of gold in general—accounting for about half of global demand, according to Enough—and activists have started pressing them to commit to using conflict-free gold. Electronics companies don't use anywhere near as much, but the PPA is partially funding a pilot project to track gold mining in eastern DRC, and Lezhnev hopes the two industries will cooperate on such projects.

As Enough has noted on its blog, eliminating conflict minerals won't solve all of the DRC's problems. But it is a piece of the puzzle American consumers have the ability to directly influence.[28] Says Bleasdale, "The process of change has to come from manufacturers and consumers. The only way any of this will stop is if local warlords realize they will make more money in peacetime—as head of a team of twelve miners, for example—than they will in war."

MADE IN CHINA

Smartphone design, engineering, and component manufacturing take place around the globe but converge in one place to produce a smartphone. Companies called contract manufacturers handle much of this final assembly work, piecing together parts built on multiple continents into a whole. The British newspaper the *Sunday Times* has described the creation of an iPhone as part transatlantic journey, part artistic fusion:

> Advanced semiconductors based on designs from British firms such as ARM . . . are shipped in from . . . Taiwan. Memory chips are imported from Korea and Japan and touchscreen and display panels and circuit boards are flown in from Korea and Taiwan, along with rare metals from Africa and Asia. [Contract] workers in China . . . [then] weave together these components from a blueprint dreamt up in Silicon Valley.[29]

Contract manufacturing in the consumer electronics industry dates back to the 1970s, when computer makers needed help assembling printed circuit boards. These days contract manufacturers build an array of consumer electronics, from laptops to game consoles to smartphones. Their purpose is to construct products for other companies at low cost and high efficiency, using the most advanced manufacturing processes. Along with design and manufacturing, contract manufacturers may also provide product distribution, warranty, and repair services to their customers.

If you have an HTC, LG, or Samsung smartphone, the person who assembled it may have been directly employed by that company. But if you have an iPhone, a BlackBerry, or a Motorola phone, your phone was likely made by someone working for a contract manufacturer. HTC, LG, and Samsung all operate phone assembly plants in their home countries. This is feasible because the cost of labor in Korea and Taiwan is relatively low. But rising wages and cost-cutting measures are threatening this model. HTC recently hired Chinese and Taiwanese contract manufacturers to produce some of its lower-cost smartphones. Samsung builds some of its midrange smartphones in China and manufactures tens of millions of phones a year, including many of its high-end smartphones, in Vietnam.

Most other smartphone makers already outsource these tasks. Hiring a contract manufacturer enables phone makers to reduce assembly costs, increase production rates, and focus their attention on other parts of their business. Outsourcing smartphone manufacturing is a key part of BlackBerry's survival plan under new CEO John Chen. In late 2013 BlackBerry announced that it had signed a five-year strategic partnership with Hon Hai Precision Industry, a Taiwan-based contract manufacturer better known as Foxconn. The partnership calls for Foxconn to jointly develop and manufacture new BlackBerry devices and manage the inventory associated with them—a task BlackBerry had struggled with, costing it about $2.5 billion in unsold phone inventory in 2013 alone. Chen has said the partnership will let BlackBerry focus on design and software development while "leveraging Foxconn's scale and efficiency" to "compete more effectively."[30]

With the exception of the 5C, most iPhones are produced by Foxconn. Geoffrey Crothall, the communications director for China Labour Bulletin, a nonprofit advocacy group dedicated to Chinese workers' rights, says Apple is yoked to Foxconn by necessity. "Foxconn is the only supplier that can really produce the volumes Apple needs in short periods of time with the quality control that Apple demands," says Crothall.

Foxconn can accomplish this because it is China's largest private employer, with more than 1 million workers. It also operates an industrial complex often described as the world's largest factory, a Shenzhen facility most people call Foxconn City because the almost 1-square-mile campus encompasses more than a dozen buildings, including dormitories that look like high-rise apartments. A typical factory in southern China employs 500 to 5,000 people. Foxconn City employs more than 135,000 people, and nearly 40 percent of them reside in Foxconn's dorms.

Foxconn City produced iPhones until relatively recently. That meant employees took iPhone displays, circuit boards, and other components and assembled them to create working devices. Line workers also loaded software onto the phones, tested them for bugs, and boxed them up with accessories to ship to consumers around the world. These days Foxconn City makes iPad tablets, Macbook laptops, and non-Apple gadgets, while iPhones are assembled in smaller Foxconn factories elsewhere in Shenzhen and other Chinese cities.

Foxconn City in Shenzhen, China (*Elizabeth Woyke*)

When people talk about the human costs of producing smartphones, they are almost always referring to Foxconn's work for Apple. This is true, even though most smartphone makers use some form of low-paid factory help. Labor-rights advocates say they single out Apple because it rakes in particularly high profits while pressing Foxconn to keep costs as low as possible. Like Apple's component suppliers, Foxconn's gross margins are relatively thin, currently around 13 percent. And Foxconn's iPhone workers earn about $2 an hour.

Activists also blame Apple's dramatic phone launches and just-in-time production strategy for elevating worker stress. "Apple will advertise a new phone and say it's available in a few weeks," says Li Qiang, executive director of China Labor Watch, a New York City–based nonprofit focused on Chinese labor issues. "They don't give factories enough time to find workers or workers enough time to produce." Each new iPhone debuts in more countries than the previous iPhone. The iPhone 4 launched with availability in five countries, the iPhone 4S in seven countries, the iPhone 5 in nine countries, and the iPhone 5S in eleven countries. Apple follows up this initial rollout with several more waves of releases. New iPhones are usually available in 100 countries within three months of their original launch.

Prior to each launch, Apple prepares a stockpile of new iPhones to meet preorder, launch-day, and early demand. Creating that stockpile can be stressful and, after it runs out, iPhone production can be frantic. Compounding worker strain, iPhones keep getting more sophisticated, and thus harder to build. Making an iPhone 4S required 141 work steps.[31] Foxconn executives have said the iPhone 5S takes longer to assemble than the iPhone 5, and the iPhone 5 was more difficult than the iPhone 4S, indicating the number of steps has increased since then.

Foxconn work hours fluctuate according to department and Apple's production schedule. Foxconn started reducing work hours in 2013, but prior to then, off-season shifts could run from 8:00 A.M. to 8:00 P.M. six days a week, with one-hour breaks each for lunch and dinner. And in the run-up to an iPhone launch workers could clock as many as 80 overtime hours a month, because they worked through some meals and the weekend. Since iPhones are introduced in late September, the busy season starts a few weeks earlier. Workers can get swamped for months, because iPhone demand stays high through December, due to the holiday shopping season, and workers may also have to work overtime in January and February to fill in for colleagues who go home for Chinese New Year.

Labor-rights advocates say life is grim for Foxconn workers whether it's peak iPhone production season or not. (Foxconn did not respond to requests for comment.) Debby Chan Sze Wan, the former head of Students & Scholars Against Corporate Misbehavior (SACOM), a Hong Kong–based workers' rights NGO, says Foxconn treats its workers like machines. "They have to stand [for as much as] twelve hours a day and keep doing monotonous tasks," she explains. "They can't talk to each other. If they slow down or make mistakes, they will be scolded. It is a cold and inhumane working environment."

Even Foxconn workers' off hours can be bleak. According to activists, the company's huge dormitories are filled with migrants who speak disparate Chinese dialects. Workers are randomly placed as many as eight to a room and rarely see each other, because they are assigned different shifts in different departments. Sometimes roommates don't even know each other's names. Foxconn says it began outsourcing the maintenance and operations of its on-campus housing to property management companies in 2010 and that its workers now have "more flexibility to choose" who they want to live with and their number of roommates, which can range from three to seven (for four to eight people total in a single residence.) Foxconn also says it has been promoting off-campus living since 2010, especially at its newer factories, in order to encourage employees to "immerse into [their] local communities" and "create a social network outside of work." Nevertheless, Li and Chan want Apple to increase the amount it pays Foxconn so Foxconn can raise wages and upgrade its living facilities. They say Apple should either absorb the cost or pass it along to consumers through higher iPhone prices. Li says, "If you talk to factories in China, they say they're fine with making changes, but that you need to talk to the brand [commissioning their work] about giving them more money."

In 2010, Foxconn's harsh culture provoked a series of worker suicides that shocked the world. Between January and November 2010, 18 Foxconn workers attempted suicide, many by jumping off Foxconn buildings. Fourteen died. One of the survivors, Tian Yu, said Foxconn's management had made her depressed and angry.

Yu recounted the events from her hospital bed in a 2010 video interview with SACOM. She said she was 17 years old and had worked at Foxconn City for about six weeks. It was her first time living outside her home village, and her first factory job. Her assignment was to check screens for scratches. She looked at screens from 7:40 A.M. to 7:40 P.M. and had one day off every second week. She didn't know what she was producing; she didn't even know if the screens were for phones, tablets, or computers. Her supervisor regularly yelled at her, though she says she did not make any mistakes. At the end of her first month an administrative error kept Yu from collecting her paycheck. She was told to go to another Foxconn campus an hour away to handle the issue. The other campus was smaller but still had about 130,000 workers. Yu spent several hours searching for help, but she says everyone she approached brushed her off. She returned to Foxconn City without her paycheck. The next morning, broke and despondent, she jumped from her fourth-floor dorm window. After spending 12 days in a coma, Yu was released from the hospital paralyzed from the waist down. She returned to her hometown in central China in a wheelchair. A few months later, Foxconn sent her parents a $28,000 "humanitarian payment."[32] SACOM says Yu remains at home and continues to receive rehabilitation.

The events that precipitated most of the other Foxconn suicides are unknown, but Yu's story appears to be representative. Migrant workers from poor, rural areas of China make up 99 percent of Foxconn City's workforce. This is also true of Foxconn's Guanlan factory, also located in Shenzhen, which produces iPhones. The average age of all Foxconn workers is 23; production line workers are younger overall, and can be as young as 16. All of the Foxconn workers who attempted suicide in 2010 were migrants and nearly all were aged 17 to 25. Many had never had a job or lived on their own before. In 2014, a group of Chinese sociologists, writing in the academic journal *Current Sociology*, characterized the Foxconn workers' suicides as "an act of frustration" about their "marginalized status and meaningless life" and a "means of accusation" intended as a wake-up call to China and the world.[33]

Foxconn took a few months to react to the suicides. CEO Terry Gou later told *Bloomberg Businessweek* that he didn't consider the first few suicides to be "a serious problem" because so many people work at Foxconn City that it did not appear to be an alarming trend.[34] In May 2010, about two months after Yu jumped from her dorm, Foxconn established mental-health resources for troubled workers, including an emergency hotline and a 24-hour counseling center. To discourage jumping, Foxconn installed 15-foot-wide nets under Foxconn City building roofs and locked windows shut. In June 2010, after ten of its workers had committed suicide, Foxconn raised wages for its production-line workers 30 percent, to 1,200 yuan ($198) per month. A few days later Foxconn announced plans to raise wages again, by nearly 70 percent, from 1,200 yuan to 2,000 yuan ($331) for workers who passed a three-month performance test.

Apple's public response to the suicides was subdued. In May 2010, the company issued a written statement saying it was "saddened and upset" by the Foxconn suicides.[35] During an appearance at a June 2010

AllThingsD conference, then CEO Steve Jobs called the suicides "very troubling" but assured the audience, "Foxconn is not a sweatshop." [36] Apple was more active behind the scenes. That month the company sent several executives and two suicide prevention specialists to its Foxconn factories. In July 2010 Apple commissioned a broader team of experts to do an independent review of Foxconn's management. The team suggested Foxconn improve training for its hotline and counseling center staff and to monitor conditions more closely "to ensure effectiveness." [37]

Consumers may have been disturbed by the news coming out of China, but they didn't shun Apple or iPhones—at least, not many did. Apple enjoyed record revenues and iPhone sales throughout Foxconn's darkest period of suicides (from late 2009 to early 2011). The iPhone 4 launched in late June 2010, just a few weeks after a spike in Foxconn suicide attempts, yet the phone sold enough units (1.7 million) during its first three days for Apple to declare it its most successful product launch ever.

Beyond Apple and Foxconn

Some people in the smartphone industry find the media and activist fixation on Apple and Foxconn excessive. iFixit CEO Kyle Wiens says other Chinese factories have far worse labor conditions but are ignored because they don't make household-name consumer goods. "Everyone's focused on Foxconn," says Wiens, "but they have some of the best factories and dorms and pay in China."

China Labor Watch and SACOM acknowledge that questionable labor practices in the smartphone industry extend beyond Apple and Foxconn, and abuses are often more egregious at other companies. The two groups have probed several Apple and Samsung partners and suppliers through undercover inspections and worker interviews. Compared to other Apple contract manufacturers, "Foxconn wages are slightly higher, its overtime slightly lower, and its safety training meets legal stipulations," says China Labor Watch program coordinator Kevin Slaten.

Both China Labor Watch and SACOM recently investigated

Pegatron, a Taiwan-based contract manufacturer that assembles the iPhone 5C, as well as other gadgets. China Labor Watch's researchers said they found multiple labor rights violations in three of Pegatron's Chinese factories, one of which produces the iPhone 5C. The group described the conditions, including low wages, excessive overtime, and dorm rooms packed with up to 12 people, as "even worse" than those at Foxconn,[38] and accused Pegatron of relying on labor violations to increase its "competitive advantage" in order to win more contract manufacturing work from Apple.[39] Pegatron said it would investigate the allegations.

Apple and Pegatron faced renewed scrutiny a few months later, when news broke that five Pegatron workers had died over the course of 2013. The most shocking case involved a 15-year-old who used his older cousin's ID to obtain a job, then succumbed to pneumonia a month later. China Labor Watch blamed Pegatron's long hours, but Apple said it had investigated the factory and found the deaths were caused by medical conditions unrelated to Pegatron.

SACOM recently bestowed the disparaging name "Foxconn the second" on a factory that processes glass for iPhone and Samsung phone touchscreens.[40] A Hong Kong–based producer called Biel Crystal Manufacturing owns the plant, which is located near Shenzhen and specializes in the dusty, dangerous work of grinding smartphone touchscreens to size and cutting tiny pieces of glass to cover smartphone cameras. SACOM said Biel Crystal workers have developed health problems from inhaling glass-polishing chemicals and glass dust particles and are denied fair compensation if they are injured on the job; five workers have reportedly committed suicide since 2011. In response, Biel Crystal management agreed to a few reforms, including increased assistance for injured workers.

Samsung also has a checkered labor record. China Labor Watch investigated 15 Samsung-affiliated, China-based factories in 2012 and 2013 and says it found underage workers (below the legal Chinese working age of 16) and salaries and working conditions "well below" legal standards.[41] While Samsung makes most of its own products, it outsources some work to suppliers. The watchdog group says Samsung's supplier factories tended to be worse than its owned-

and-operated ones, but all the Samsung-related facilities checked "mistreat[ed] labor in one way or another."[42]

In response to the allegations Samsung conducted on-site inspections of all of its Chinese suppliers. The company said it found no evidence of child labor but did uncover "inadequate management" and "potentially unsafe" health and workplace policies.[43] In late 2012, Samsung began implementing new hiring policies, safety measures, and overtime practices in China, but a few months later a worker suicide at one of its suppliers cast a pall over its reforms. Li says, "All the factories we've investigated in the past improve somewhat [due to public pressure], but if we don't go back and reinvestigate in a year or two, they might just go back [to their earlier conditions]." In 2013, after exposing a number of labor violations at a China-based factory that produces phone cases and screens for Samsung—and, bizarrely, requires many of its workers to work barefoot—China Labor Watch concluded, "Labor abuse has persisted at Samsung supplier factories."[44]

Labor tensions are expected to flare up in other countries as smartphone makers and contract manufacturers shift work to Southeast Asia and Latin America to keep costs low and profits high. In the 1990s, when gadget makers rushed to outsource manufacturing to China, labor was "so compellingly cheap that all other factors were outweighed," says Aronow. But Chinese wages more than tripled between 2000 and 2010, according to the International Labour Organization (ILO) and are expected to double over the next ten years. "China is not necessarily number one anymore, if you want the lowest cost," says Aronow. Both Samsung and Foxconn operate smartphone factories in Brazil and their authoritarian management style has triggered local resistance ranging from worker protests to government investigations and lawsuits.

Lingering Problems

China Labor Watch and SACOM remain focused on China—specifically, Apple and Foxconn's actions in China. In part this is because stories about Apple and Foxconn attract immediate global attention. In part it's due to limited resources. Both organizations have just a few staff members and run their operations out of humble, two-room offices. But it's also because the groups believe Foxconn still needs reform.

Liang Pui Kwan, SACOM's current leader, notes that Foxconn offset its workers' wage gains by canceling their housing and food allowances. She says workers who live on campus now must pay 110 yuan ($18) a month for housing and buy food à la carte when they eat in Foxconn's cafeteria. Foxconn says it converted these allowances to be part of its employees' take-home salaries in order to give workers "more flexibility with the way they choose to spend their subsidies." Crothall says working at Foxconn is "still very tough" and "dehumanizing." He says that workers' financial responsibilities prevent Foxconn's base salary from being a living wage: "A young, single worker with no obligations can get by on that amount, but most of these workers want to send money home or save to further their education."

Recent investigations of Foxconn's iPhone operations have focused on other forms of exploitation, including the use of temporary "dispatch" workers, who are hired through employment agencies rather than directly recruited by the company. Since these workers are technically employed by the agencies, Foxconn can quickly hire and dismiss them during its high and low seasons. Activists dislike the system, because dispatch workers are more vulnerable to labor rights violations. They say these workers are often paid lower wages and denied employee benefits, including medical insurance, compensation for work-related injuries, and pension payments, which China's Social Insurance Law requires. Employment agencies are supposed to make insurance contributions for dispatch workers but usually don't, says Chan. Foxconn says temporary employees make up less than 0.38 percent of its workforce "at any given point in time," though it declined to specify how it was defining the term "workforce" in that context. Foxconn also says that it "generally" does not use dispatch workers unless its customers "have short-term needs that require" it to do so and that when it does, it pays them the same wages as its permanent workers. In addition, Apple says it created an "ethical hiring" training program in 2013 that teaches suppliers how to manage vulnerable worker groups, including dispatch workers, without exploitation.[45]

Health effects also remain a concern for Foxconn workers. Past investigations found line workers were using potentially harmful chemicals to clean iPhone screens. Apple says it has outlawed the use of an insidious chemical, n-hexane, that causes neurological damage, but Liang says Foxconn still uses toxic chemicals. Liang alleges some workers clean iPhone casings with benzene, a liquid that can cause dizziness and unconsciousness if inhaled deeply, and that eventually affects bone marrow and blood production. Chan says workers don't know what the chemicals are because Foxconn refers to these formulas by euphemisms, such as "cleaning water" and "cutting fluid." Liang also alleges that some workers are at risk of occupational deafness, because the machinery they use to stamp raw materials to produce iPhone casings and components is loud, and they have inadequate hearing protection equipment. Apple says its suppliers are required to provide comprehensive safety training and protective gear, such as gloves, masks, goggles, and earplugs, for workers. Foxconn says company policy prevents it from commenting on the work it does for Apple, but that all Foxconn employees participate in regular occupational health and safety training and are given "the protective equipment that is required to help them safely carry out their work."

Most troubling of all, Foxconn workers are still jumping off buildings. Two fell to their deaths in 2012. Three more reportedly followed in 2013, according to activist groups.

Fair Labor at Foxconn

From early 2012 to late 2013, the Fair Labor Association (FLA) kept an eye on Foxconn. The FLA is a Washington, D.C.–based nonprofit that promotes companies' adherence to labor laws. Companies that join the FLA pay annual dues based on their revenue and agree to implement a nine-point Workplace Code of Conduct throughout their supply chains and to open their factories to "monitoring visits" or audits overseen by the FLA.[46] These generally include a walk-through inspection of working facilities, interviews with managers and workers, and a review of payroll and time-sheet records.

Apple joined the FLA in January 2012, a week after the national radio program *This American Life* aired a show featuring author and actor Mike Daisey performing a monologue that vividly accused Foxconn of employing and mistreating workers as young as twelve years old. (*This American Life* retracted the story in March 2012 after it learned Daisey had fabricated many of the most appalling parts of his story.) In February 2012, at Apple's behest, the FLA examined three Foxconn factories, including Foxconn City and a nearby facility that produces iPhones. A few weeks later, the FLA issued a report with a long list of recommended improvements related to work hours, compensation, and worker health and safety; Apple and Foxconn then created a 15-month "action plan," and the FLA spent its subsequent China visits (one trip in summer 2012 and two in 2013) tracking Foxconn's progress with the assistance of two independent China-based labor-monitoring organizations. Apple paid the FLA a fee "well into the six figures" for the audits, on top of its $250,000 annual dues.[47]

In December 2013, several months after the planned conclusion of the 2012 action plan, the FLA released its final report, which said Foxconn had implemented nearly all of the 360 "action items" except for four tasks related to reducing overtime hours.[48] The group had sought to cut Foxconn's working hours to 60 per week, which is the benchmark the FLA uses globally, and then to 40 (augmented by 36 overtime hours a month), which is the Chinese legal limit. Foxconn managed to trim its working hours enough to meet the FLA standard but failed to comply with the stricter Chinese law.

The FLA has said it expects Foxconn to keep striving to reduce work hours, but labor activists are doubtful it can comply with the legal standard, which many Chinese factories don't meet. Though Apple requires suppliers to give employees at least one day of rest per seven days of work, it asks its suppliers to cap their workweeks at 60 hours—not 49—and it permits excess hours in "unusual circumstances."[49] Foxconn says that it will "continue to work" with "third-party advisors" and customers on managing employee hours, but it declined to comment more specifically on the topic of overtime reform, such as its target number of work hours.

In general, labor advocates are skeptical of solutions that rely on audits, since the employers at factories may coach workers on what to say. To combat duplicity and retaliatory action against whistle-blowers, Apple and its partners conduct surprise audits, give workers who participate in audit interviews a hotline number for reporting mistreatment, and make follow-up calls to some workers postaudit. But activists say audits can still be manipulated, and auditors may unintentionally overlook issues workers consider important.

Instead of audits, labor-rights advocates want Foxconn to let workers form labor unions that can communicate and negotiate with company management to improve conditions. Right now, workers in Apple supplier factories can't leverage their influence as a group to demand legal and fair treatment. "There should be a day-to-day channel for workers to organize themselves," says Chan.

The idea of a collective bargaining system is simple but tough to implement, because China has an official, nationwide union called the All-China Federation of Trade Unions (ACFTU), which is a subsidiary of the Communist Party and does not permit independent, competing unions. China Labour Bulletin has described the ACFTU as an organization that "still sees itself as a third party, as a broker between labor and management, rather than as a representative of the workers."[50] Foxconn established a labor union in 2007 and began holding representative elections in 2008, but, like other Chinese unions, it is affiliated with the ACFTU and dominated by company supervisors, who are nominated for election by management. A 2013 SACOM study that included a survey of more than 600 Foxconn workers found 95 percent of respondents had never voted in a Foxconn union election, and only 17 percent had union membership cards.[51] "Foxconn has a union, but workers don't trust it or know what its function is," says Chan.

Foxconn appeared to be heading in a more grassroots direction in 2013. In February 2013, the *Financial Times* reported that Foxconn was preparing its first "genuinely representative" union elections, which would have an "anonymous ballot voting process."[52] Two days later *Xinhua*, China's official news agency, reported that Foxconn would hold elections by July 2013 and intended to "increase the number of junior employee representatives in all committees within the union."[53] China Labor Watch's Kevin Slaten says that elections may have been held in 2013, but if so they were not actually democratic.

Foxconn says its union did not hold employee representatives elections in 2013 because it only does so every three years and the previous election took place in 2011. The company declined to say whether it would hold such elections in 2014.

Lacking faith in the unions, Foxconn workers have been taking action themselves by holding spontaneous strikes and walkouts. Most of these incidents occurred in inland Chinese cities where Foxconn started building factories in 2010 to expand its labor pool. In September 2012, 2,000 Foxconn employees shattered windows at a factory in the northern Chinese city of Taiyuan. Police teams were called in and Foxconn was forced to shut the plant temporarily. The company described the disturbance as an employee fight that spiraled out of control. Activists characterized it as a worker protest exacerbated by iPhone 5 production stress. The following month more than 3,000 Foxconn workers in the central Chinese city of Zhengzhou went on strike. Activists say the workers were protesting oppressive quality control measures Foxconn imposed after consumers complained about scratches on their new iPhone 5s.

Chinese labor strikes have occupied a legal gray zone since 1982, when a revision of the country's constitution removed wording guaranteeing people's right to strike, but many workers will walk out anyway. "This is a new generation," says Crothall. "Workers know their basic rights and are no longer willing to accept that kind of tough regime."

In recent years, Foxconn has sought to reinvent itself as more than a low-cost contract manufacturer and Apple has sought to fix its labor reputation. In pursuit of faster growth and higher margins, Foxconn has recruited new partners, such as BlackBerry, and branched into new businesses, including developing Firefox OS smartphones and tablets for other companies to sell, investing in technology start-ups, and preparing to offer 4G phone service in Taiwan.

Since 2011, when Tim Cook became CEO, Apple has joined the FLA and started publishing a list of its suppliers on its website. Li views Cook as "relatively" more open to labor discussions than Steve Jobs. "In the past, Apple wouldn't publish any labor information publicly or contact NGOs," notes Li. These days, Apple is willing to speak directly with China Labor Watch. Their communication has brought about some small successes. Li says he wrote Cook in early 2013 and told him about a 15-year-old worker at Foxconn's Zhengzhou iPhone factory. A few weeks later, Apple replied and said it had sent the girl back to school with a year's tuition and some money for living expenses.

Li wishes Apple would take charge on more labor issues. "Electronics companies will talk to us for a whole day and not do anything," he says. "If Apple were to do something, it would drive reform through the entire electronics industry." Gartner analyst Aronow says, "If Apple is going to improve conditions at these large contract manufacturers, standards in general will go up for anyone who uses those services. It is for the greater good of the larger electronics industry."

5

Waste: Money and Trash

Every day, when Todd Dunphy checks his e-mail, he finds messages from people begging for assistance with their smartphone plans. One December 2013 e-mail came from a person who was paying $1,200 a year for unlimited service but only using 150 voice minutes, 25 text messages, and a scant amount of mobile data each month. The issue wasn't the person's smartphone—a Galaxy S model that the phone's owner liked—but the expense of the accompanying plan. "I just feel like I am paying so much money for services I don't use, but [I can't] find a very good [alternative] option," the owner explained. "Please help!!!"

Dunphy has been receiving these kinds of "Please help!!!" e-mails for years. He is a former Verizon Wireless salesman, and in 2007 he co-founded a mobile analytics firm called Validas to save people and consumers money on their wireless service. The Houston, Texas–based company analyzes the cellphone bills of companies, government agencies, and consumers for extraneous charges using its own software. As a result, Validas claims to now have amassed the largest collection of wireless bill information in the United States outside of the carriers.

Sifting through the data, the company says the biggest trend in spending habits is wireless waste—the gap between the service capacity people sign up for and the amount of that service they actually use. Says Dunphy, "The only way I win, as a smartphone consumer, is when I perfectly buy and use [an exact number of gigabytes], which never happens."

We all like to think we're savvy shoppers. Yet Validas's data shows that most of us spend more than we need to on wireless service month after month, year after year. Validas estimates 80 percent of American wireless subscribers are paying carriers $200 a year for excess minutes, messages, and data. On a national level that adds up to $52.8 billion of wireless waste a year. On a global level the figure could be as high as $926 billion, Dunphy says. Smartphones are the biggest source of wireless waste, because their plans are particularly expensive and complicated. According to Validas they are responsible for $45 billion, or 85 percent, of the United States' total overspending on wireless service.

As a company that aims to lower its clients' mobile costs, Validas has business incentives to call attention to wireless waste, and potentially to overstate its scope. But others in the industry agree that wireless waste is a problem. A New York–based company called Alekstra, which was founded by a former Nokia executive and provides services similar to those of Validas, has estimated that U.S. wireless waste is as high as $70 billion.[1] A 2013 *Consumer Reports* survey of more than 58,000 of the publication's subscribers found 38 percent of its respondents were using half or less than half of their monthly data allowance.[2] Logan Abbott, president of MyRatePlan.com, a comparison-shopping site for wireless plans, says the vast majority of consumers don't use all the data or minutes they're allotted.

How do we end up with these supersized smartphone plans? The short answer is that carriers keep pushing us into pricier plans to maximize their profits. The carriers wouldn't phrase it that way, but industry analysts say carriers are wringing more money out of their subscribers to counteract slowing growth. For years carriers grew by just signing up people who were brand-new to wireless service, but now virtually every American who wants and can afford a cellphone has one. Carriers can still boost their revenues by upgrading subscribers from basic cellphones to smartphones, which require more expensive service plans, but that group is dwindling, too. More than 65 percent of Americans already have smartphones.

Carriers are now primarily concerned with maintaining or increasing profitability, which, until recently, has meant cutting operational costs and raising rates on their smartphone subscribers. Smartphone users are obvious targets. Just by purchasing smartphones users indicate they are more capable of spending, or at least more likely to spend, than basic cellphone users.

THE PRICE OF MOBILE DATA

When estimating the cost of smartphone ownership, many consumers focus on the upfront cost of new smartphones rather than on the price of the accompanying plans. But the devices cost much less than the plans, which are a recurring expense.

Carriers sell two basic types of service plans: postpaid and prepaid. Postpaid plans are contractual agreements between subscribers and carriers, with carriers billing subscribers for service at the end of each month. Carriers traditionally gave postpaid subscribers deep subsidies on new phones. Since subsidies average around $400, and carriers typically price new smartphones somewhere between $0 and $250, carriers actually lost money each time they sold a subsidized smartphone. To ensure they recoup their costs and turn a profit, carriers make these subscribers agree to purchase service from them for a set period of time, usually two years.

With prepaid plans, customers pay for service in advance, by either purchasing credits for set allotments of voice minutes, text messages, and data bytes or prepaying for a bundle of services per day or month. Prepaid users also buy their phones on their own. Since carriers don't take on any financial risk with prepaid plans, users aren't required to sign contracts.

In many regions, such as Africa and Latin America, prepaid plans are more prevalent, but in the United States postpaid plans are more common, especially for smartphone owners. About two thirds of U.S. cellphone users are on postpaid plans and one third are on prepaid plans. To carriers, postpaid customers are more valuable, because they generate more revenue.

The future of the carrier business lies in mobile data. Carriers in the United States and a few other countries, such as Japan, already make more money from data than voice services and most U.S. carriers now automatically include unlimited voice in their smartphone plans. As the *Wall Street Journal* pointed out in October 2013, "The carriers aren't making this change because they think customers should just have unlimited calls for free. . . . Focusing on data and giving everything else away seemingly as a freebie actually centers the companies on their main source of revenue growth."[3] Mobile data is a $90 billion market in the United States alone; globally, it is worth about $400 billion.[4]

Smartphones and data plans are inextricably linked. For the most part, carriers won't let consumers buy smartphones without signing up for mobile data plans. Asymco analyst Horace Dediu calls the iPhone a "mobile broadband salesman." Carriers "hire" the iPhone to sell mobile data plans to consumers, and the phone's job is to move people from $50-a-month plans to $100-a-month plans by transitioning them from basic to smartphones. The prices that carriers pay Apple are essentially sales commissions for accomplishing this job. "Apple gets a huge commission in the form of a price premium and subsidy on the iPhone," Dediu explains. "It's the carriers saying, 'Thanks, you've helped us attract more customers to our most profitable service plans.' "

Those service plans can change frequently, in both composition and price. Carriers are comfortable raising rates, because they know consumers are dependent on their smartphones and apps and many are locked into multiyear contracts. "Smartphones are close to a life necessity now," says Dunphy. "It's become 'food, water, shelter,' and 'Do I have my phone with me?,' and the carriers know that."

One common carrier move is to kill unlimited data plans. A few years before the iPhone was unveiled, carriers started rolling out 3G technology, which increased their capacity to support mobile data usage. To spur consumers to access the mobile Internet on their phones, they introduced unlimited plans at affordable rates. But iPhone, Android, and other smartphone users started sucking up more bandwidth than anticipated. "Unlimited plans became the operators' arch enemy," says Jake Saunders, ABI Research's vice president of forecasting. "Operators were blindsided by the small percentage of users who took those plans to the ultimate limit." In response, AT&T and Verizon adopted tiered pricing to safeguard their networks and profits: subscribers select the amount, or "tier," of voice minutes, text messages, and data they believe they will use per month. Subscribers who consume more than their allotments are charged a high rate for extra data usage.

Today only 12 percent of mobile subscribers around the world have access to unlimited "all you can eat" data plans, down from 15 percent in mid-2013, according to ABI Research. Tiered, wage-based plans are much more prevalent—they are available to 64 percent of mobile subscribers globally. In fact, unlimited data has become so rare that Sprint launched an unlimited guarantee policy in 2013 and is touting it as a major asset for its customers. But how long can Sprint buck the industry trend toward tiered plans? Sprint has fewer subscribers than AT&T and Verizon, so it has more bandwidth to support unlimited plans. Following its 2013 merger with the Japanese carrier SoftBank (Japan's number-three carrier and now owner of about 80 percent of Sprint), the formerly cash-strapped Sprint also now has funds it can use for network upgrades and maintenance. Deepa Karthikeyan, an analyst for the market research company Current Analysis, says Sprint has to "pull out all stops to create a strong differentiator against the leading carriers," for the next couple of years. But if it "gains critical traction" in the marketplace and no longer needs such enticements to lure consumers away from its rivals, she thinks Sprint may stop offering its unlimited guarantee to new customers. In a sign that Sprint is seeing network congestion, it recently warned its top 5 percent of data users that it may start reducing their mobile throughput speeds "during heavy usage times" to keep connections fast for other customers.[5]

When AT&T first adopted tiered pricing, the company said the change would "make the mobile Internet more affordable to more people,"[5] because its lowest data tiers cost less than the unlimited plans they were replacing. That was true, but the catch is that bandwidth consumption is escalating as smartphones gain features, a wider range of high-definition media launch, and data connections get faster. Karthikeyan says that "the overall smartphone experience" keeps improving for consumers, but they are also spending more on tiered pricing plans. Subscribers may start out paying relatively little in the lowest tiers, but they quickly move up to more expensive ones, because they need more data and they are charged overage fees if they exceed data caps.

Carriers have also been increasing data prices. Consider the short history of AT&T's data plans for the iPhone. The iPhone launched in 2007 with a $20 unlimited data plan. A year later, when the iPhone 3G debuted, AT&T upped the price of unlimited data to $30. In June 2010, AT&T switched to tiered pricing, which meant new iPhone 4 buyers had to choose between paying $15 for 200 MB or $25 for 2 GB. Today an individual AT&T iPhone user can still pay $20 for data but will get only 300 MB, and buying 2 GB costs $40. Verizon's plans have followed a similar trajectory and are even more expensive. Until July 2011, unlimited data on Verizon cost $30. After it moved to tiered plans, $30 bought 2 GB. As of May 2014, $30 will buy 500 MB on Verizon, and 2 GB costs $50.*

New types of mobile data plans can cost consumers even more. Multidevice shared-data plans allow subscribers to use a set amount of 4G LTE data among different types of devices and pay only one monthly fee. These plans look pragmatic on paper, since the costs and data can be divvied up many ways, including among family members. But analysts say the plans can easily accelerate consumers' data consumption and costs, on top of which they must pay a monthly line access fee for each device (independent of data usage). The result is that these plans end up being 10 percent to 15 percent more expensive than other mobile data options, according to Saunders.

Multidevice plans are also more effective at locking in subscribers, since it's more complicated to switch service and carriers when multiple devices and people are involved. Ralph de la Vega, the president and CEO of AT&T's wireless business, has described the company's Mobile Share plans as "a good plan for us from a loyalty point of view."[7] AT&T showed its enthusiasm for multidevice shared-data plans in October 2013 when it dropped its regular tiered plans and began requiring new subscribers to sign up for Mobile Share plans. Only 9 percent of global mobile subscribers currently have access to multi-

---

*Though prices will change for any service cited in this chapter, practices are likely to remain the same in the foreseeable future.

device shared-data plans but that number has nearly doubled since May 2013, showing that carriers are quickly adopting this newer type of plan. Saunders expects more carriers will follow, noting, "Operators realize . . . they can get more money out of the user" with them.

Consumers would make better choices if they understood what they were buying. In July 2013, AT&T, Sprint, T-Mobile, and Verizon offered a total of nearly 700 combinations of smartphone plans, according to a *Wall Street Journal* analysis.[8] Which plan is best? The only honest answer is: it depends. Dunphy advises consumers to buy voice minutes, text messages, and data the same way they buy milk. He means that people should take more care to buy only what they need, since anything left over will be thrown away (or taken away by the carriers, in the case of wireless plans). "Smartphone plans are supercomplicated, with lots of variations," admits Dunphy. "But we're trying to get people to think about the way they buy other things in life and apply that thinking to their wireless plans."

The problem is that mobile data is not sold the way most products are. In order to select a data tier, consumers must estimate how much bandwidth they will use in a month. But most people have no idea how to quantify mobile data, which can fluctuate due to a number of variables. An e-mail with attachments takes up more bandwidth than a text-only e-mail. Browsing graphic-heavy, video-enabled websites requires more data than surfing simple sites. Streaming video over 4G is more bandwidth-intensive than 3G video. Streaming high-definition video can eat up more than twice the amount of data as standard video.

Even carriers disagree about the amount of data a particular activity will consume. Verizon says downloading a song will eat up 7 MB; AT&T says 4 MB. Verizon says sending a text-only e-mail should take 10 KB; AT&T and Sprint say 20 KB. AT&T estimates video streaming will consume 2 MB a minute; Sprint and Verizon say it's more than double that, on average. The three companies also list different rates for music streaming (500 KB or 1 MB per minute) and Web surfing (either 256 KB, 400 KB, or 500 KB per page). These are just estimates, which will vary by smartphone model and operating system, and which reflect technological differences between the networks, such as download and upload speeds. Nevertheless, the variance between the carriers' numbers may increase consumer confusion about data rates and consumption.

Uncertainty breeds risk aversion. People overspend because they are anxious they will exceed their voice and data quotas. Most consumers would rather oversubscribe and know they'll stay within their plans' bounds than pay less and have to worry about penalties. Carriers may stoke those fears to make more money. "Sales reps will say, 'Go with a fifteen GB plan; you have a teenager, you need that,' [even if you don't]," says Dunphy. He blames the industry's sales commission system, which he claims has not changed much since he worked in a New Jersey Verizon Wireless store in the early 2000s. He says: "Every single thing [your carrier sells you], someone is getting a commission on. When a sales rep sells you a higher plan when you should be on a lower plan, it's because he's getting paid based on the size of that plan. The system is built like that. Your bill is directly tied to commission."

Carrier Economics

Carriers say they need to charge more because their upfront expenses keep mounting. Bandwidth alone is a huge cost. Cisco, the data-networking giant, says smartphones devour 48 times more mobile data than basic phones, because people use them more like computers—surfing the Web, uploading photos to Facebook and Instagram, and streaming music from Pandora. AT&T says mobile data traffic on its network increased more than 30,000 percent between 2007, the year it launched the iPhone, and 2012. Ericsson's latest Mobility Report, which it publishes annually based on data from its mobile networking business, forecasts that global smartphone traffic will grow 10 times between 2013 and 2019, when it will reach 10 exabytes, or 10 billion gigabytes per month.[9]

To keep pace, carriers are spending billions on wireless spectrum—a broad swath of radio frequencies that can carry a variety of communications information, from radio to TV to smartphone signals. Spectrum is a finite resource, so demand far outweighs supply. The government is responsible for allocating spectrum—to carriers, broadcasters, the military, police, emergency medical technicians, and other groups—and licenses for its usage can be purchased in government-hosted auctions or in company-to-company deals. These can cost millions or even billions of dollars. In 2012, Verizon paid four cable companies, including Comcast and Time Warner Cable, $3.9 billion for spectrum with nearly national coverage. In 2013, Sprint paid U.S. Cellular, the country's number five carrier, $480 million for spectrum in the Midwest, and AT&T paid Verizon $1.9 billion for spectrum Verizon was not using in 18 states. In 2014, T-Mobile bought some more of Verizon's unused spectrum, in several places across the country, for $2.4 billion in cash and spectrum swaps. Carriers pay these huge sums because spectrum gives them capacity to support their users; running low leads to dropped calls and data slowdowns on their networks. People in the wireless industry often call spectrum the lifeblood of mobile communications.

The spectrum shortage is driving expensive mergers and acquisitions. In 2013, T-Mobile spent $1.5 billion to merge with the midsize carrier MetroPCS and Sprint paid $3.5 billion to acquire the portion (approximately 49 percent) of the wireless broadband provider Clearwire it didn't already own. In 2014 AT&T purchased the midsize carrier Leap Communications for $1.2 billion. The buyers gained new subscribers and other assets, but a need for spectrum drove the deals. As the tech news site Gigaom put it, "The big four [carriers] are becoming chop shops, buying up smaller players and stripping them to get at their airwaves." [10]

To stay competitive, carriers have also spent billions upgrading their networks to LTE, to be faster and more efficient. Even after a carrier has rolled out an LTE network nationally, it still needs to continually expand capacity because speeds dip as more smartphones move onto the network. Faster networks also enable people to do more on their smartphones, which further increases traffic on networks, necessitating even more equipment upgrades and spectrum purchases. Cisco estimates that a 4G-connected smartphone generates more than three times more traffic than a smartphone on a slower connection. Verizon has said its LTE users consume more than twice the amount of mobile data as its 3G customers. U.S. carriers spent $11 billion on LTE equipment and software in 2013, according to the technology research firm iGR, and will spend a cumulative $38 billion on LTE equipment and installation, as well as $57 billion to keep their networks running, through 2018.

Billion-dollar infrastructure investments have divided carriers into classes. AT&T and Verizon, which together hold about 67 percent of the U.S. wireless market, have more than enough subscribers and revenue from subscriber fees to offset their costs. Verizon reported a healthy 49.5 percent wireless service margin in 2013, and AT&T's wireless margin for the year was 41 percent. For smaller carriers, such as Sprint and T-Mobile, which hold 16 percent and 14 percent of the U.S. wireless market, respectively, "the picture is much bleaker," says Greg Linden, a researcher at the University of California, Irvine, who has studied the economics of the smartphone industry.

Sprint and T-Mobile executives have publicly argued that they should be allowed to combine forces to narrow AT&T and Verizon's lead. In separate February 2014 calls with analysts, T-Mobile CEO John Legere said, "Over time, this industry is ripe for the impact of further consolidation," and Sprint CEO Dan Hesse asserted that "further consolidation in the U.S. wireless industry"—outside of AT&T and Verizon—would improve the wireless industry's "competitive dynamic" and be "better for the country and better for consumers." [11] But a Sprint–T-Mobile merger would likely rouse government resistance, if not opposition, due to antitrust concerns. Both the FCC and the Justice Department have stated that they believe the U.S. wireless industry needs four large carriers to remain competitive.

One issue that unites all carriers is their frustration with popular, bandwidth-heavy services such as Netflix and YouTube. Carriers call these services over-the-top, or OTT, because they run on top of their networks and cause congestion without paying for the data traffic they create. In October 2013, Google said that 40 percent of YouTube's traffic comes from mobile devices, including smartphones, up from 25 percent in 2012 and 6 percent in 2011. Verizon recently said video accounted for 50 percent of its wireless traffic and will rise to 66 percent by 2017. Michael O'Hara, chief marketing officer of the GSMA industry group of carriers and related companies, says, "There's a disconnect between the [network] investment required by operators and [the fact that] the revenues generated [from the networks] are predominantly flowing to content companies."

Carriers have struggled for years with this conundrum. In some areas of the world, they have struck deals with the OTT service providers clogging their networks. Google has paid Orange, the France-based carrier, an undisclosed fee since 2012 for the (mostly YouTube-related) traffic it sends across Orange's network.[12] The idea that payments for mobile services can and should be differentiated could pave the way for a number of future pricing changes, depending on the local "network neutrality" rules. (Those rules are currently in flux in the United States.) Dunphy thinks American smartphone bills will eventually resemble cable bills, with OTT services priced separately and bundled into packages similar to the way cable companies charge for channels. "Carriers could have a package in which you get 2 GB and all your Facebook and Twitter [access] included," explains Dunphy. "To get YouTube, they might make you pick the highest plan."

THE UN-CARRIER MOVEMENT

It's easy to criticize carriers—and many people do. Carriers' customer satisfaction and corporate reputation ratings are usually dismal. Telecommunications companies ranked number 15 out of 17 industries in a 2013 survey commissioned by the Reputation Institute, a New York City–based advisory firm that provides "reputation consulting."[13] Its rating suggests American consumers view telecom companies as even less reputable than the pharmaceutical, utility, airline, and energy industries, and only more reputable than banks and "diversified" financial companies. Some carriers are aware of this. In a speech at an industry conference in 2012, Sprint CEO Dan Hesse lamented the situation, saying, "[E]ven the cable and oil industries rate higher with consumers than we do."[14] Many American smartphone users, however, would probably sympathize more with cellphone inventor Martin Cooper when he says: "I spent my whole life battling with carriers. They come out of a monopoly environment and try to tell you what you want. The motivation of every carrier is to get customers to spend as much money on their networks as possible. We've built a system where the motivations of carriers are counter to the public interest."

There are a couple of areas in which the Big Four carriers—AT&T, Sprint, T-Mobile, and Verizon—are slowly becoming more consumer-friendly: phone subsidies and upgrade policies. On the one hand, carriers want customers to upgrade their phones, because upgrades keep customers locked into contracts and loyal. On the other hand, carriers don't like paying subsidies for phone upgrades. They didn't mind fronting that cost when a lot of their customers were switching from basic to smartphones and from 3G smartphones to 4G smartphones, because those changes led to more data usage, giving carriers a revenue boost. But now that consumers are upgrading from 4G smartphones to other 4G smartphones, carriers get smaller payoffs from footing subsidies. In December 2013, AT&T CEO Randall Stephenson told an investor conference that the phone subsidy model "has to change" because carriers are now in "maintenance mode" and "can't afford to subsidize devices [aggressively]" anymore, according to CNET.[15] ABI says subsidies are carriers' single largest cost over the lifetime of a subscriber's contract, eating up 68 percent of the revenue derived from a typical 24-month service agreement.

Between 2011 and 2013, carriers have rolled out a number of changes to reduce their subsidy expenses and improve their margins, including altering their discount, fee, and eligibility policies to discourage upgrades. In January 2011 Verizon canceled its long-standing New Every Two program, which let subscribers shave $30 to $100 off the already discounted price of a new phone if they had completed at least 20 months of their two-year contracts. In 2011 and 2012, AT&T, Sprint, and Verizon increased the upgrade fees they charged subscribers who bought new phones on subsidy. Those fees now hover between $30 and $36, about double what they were in 2011. In 2013, AT&T and Verizon lengthened the amount of contract time subscribers must complete to qualify for subsidized upgrades (from 20 months to 24). To thwart subscribers from acquiring new phones on subsidy and leaving early, AT&T, Sprint, and Verizon also raised the early termination fees (ETF) they charge subscribers who don't fulfill their two-year service contracts. ETFs for smartphone users have doubled since 2009, from about $175 to as much as $350.

In 2013, after years of instituting higher fees, longer upgrade cycles, and stiffer penalties, the wireless industry started emphasizing lower prices and greater consumer choice. T-Mobile has been one catalyst for change. Under CEO John Legere, who assumed his post in September 2012, T-Mobile is trying to improve its perennially fourth-place prospects by branding itself as a maverick "un-carrier." As an "un-carrier," T-Mobile purports to "break the rules" of the "out-of-touch wireless carrier club" and take the customer's side.[16] As Dunphy puts it, "T-Mobile basically said, 'What does everyone hate about wireless service in the U.S.? Let's say we agree with all those things and introduce new policies to combat them.' "

T-Mobile kicked off its un-carrier strategy by slashing prices and eliminating conventional subscriber contracts and subsidies. T-Mobile's Simple Choice plans, which launched in March 2013, sell smartphones at full price and let customers pay off the cost through monthly installments, usually over two years. The setup is similar to 0 percent financing for a car. T-Mobile customers pay for their voice, text message, and data service via a separate monthly fee.

The Simple Choice plans relieve T-Mobile of the burden of subsidizing new smartphones while giving consumers a manageable way to pay down their phone expenses over time. The setup also saves consumers money. Most people don't realize it, but approximately $20 of a traditional postpaid monthly smartphone bill compensates the carrier for its phone subsidy. That $20 doesn't show up as a separate line item on subscribers' bills, but it is included in their overall monthly fees. Since the subsidy reimbursement is built into subscribers' monthly bills, carriers continue charging it even after they recoup their subsidy costs, which usually happens between the twentieth and the twenty-second month of a two-year contract. "Carriers end up making an extra $20 a month on your contract," says Sina Khanifar, an activist who advocates for consumer rights in tech matters. "It's a big financial game that relies on consumers not understanding the economics."

Since T-Mobile charges lower service rates than its competitors, its customers save money immediately and increase their savings the longer they hold on to their phones. T-Mobile's strategy also lets customers see how much their phone really costs, how much their service really costs, and how close they are to paying off their phones. David Pogue, in a *New York Times* review, said T-Mobile's policies were "much more fair, transparent, [and] logical" than other carriers' policies.[17]

Critics say T-Mobile's Simple Choice plans are less radical than the carrier professes. The company calls the plans "no contract" and "contract-less," because customers can leave any time they want. But customers must pay the remaining balance on their phones when they leave. As CNET has noted, the device payment agreement essentially acts like an early termination fee, which means, "For many, the monthly installments constitute another contract, just one that's worded in a different way."[18] In April 2013, the attorney general in Washington state, where T-Mobile is based, accused the company of deceptive advertising for "promis[ing] consumers no annual contracts while carrying hidden charges for early termination of phone plans."[19] In the resulting court-ordered agreement T-Mobile agreed to state "the true cost" of its phones more clearly in its ads.[20]

Consumers have largely embraced T-Mobile's un-carrier changes, which also included the introduction of an early-upgrade plan for people who want to change their smartphones frequently, and free, unlimited data for Simple Choice subscribers in more than 100 foreign countries, and eliminating overage fees for customers who exceed their plan limits. In 2013, the carrier began signing up more new postpaid customers than it lost for the first time in four years. In a January 2014 note to investors, Barclays Capital telecom analyst Amir Rozwadowski said T-Mobile's gains, which amounted to more than 2 million T-Mobile-branded postpaid (net) subscribers in 2013, were "better than expected" and "clearly indicate that its un-carrier strategy continues to resonate with subscribers."* [21]

Competitors reacted by lowering prices, increasing data allotments, and introducing more flexible plans that let consumers upgrade phones faster and opt out of long-term service contracts provided they paid full price for their smartphones or acquired them on their own. To emphasize the changes, AT&T and Verizon renamed their plans— to Mobile Share Value and More Everything, respectively. While FierceWireless, an influential wireless industry newsletter and news site, cautioned against calling the changes a "price war," [22] because the discounts weren't drastic and the consumers who saved money on service fees were also paying more for their smartphones, a *New York Times* article said T-Mobile's success had put "the other carriers in somewhat of a defensive crouch." [23]

Though T-Mobile's un-carrier policies are a conspicuous new influence in the wireless industry, they are not the only force driving change. The demographics and economics of the maturing U.S. smartphone industry are also prompting these moves. As Rozwadowski has written, "Smartphone penetration has clearly surpassed the steep part of the adoption curve for the overall market with the remaining growth coming from . . . more price sensitive subscribers." [24]

Amina Fazlullah, policy director at the Benton Foundation, a Washington, D.C.–based nonprofit that advocates for the public interest use of communications technologies, says U.S. carriers are realizing they need to offer lower prices and more consumer-friendly provisions to keep growing. Fazlullah says: "Because T-Mobile and [to some extent] Sprint have taken more aggressive stances on trying to compete on a consumer-friendly scale to attract consumers, we've seen AT&T and Verizon try to soften their stance and try different ways to bring people in, too. A piece of this is driven by [industry] competition, but also the reality that this is the last group of untapped

---

*T-Mobile's flurry of promotions did cost it, though, and was a major reason the carrier reported net income losses for the second, third, and fourth quarters of 2013 and in early 2014.

[smartphone] consumers. Every carrier knows the last frontier for [smartphone] penetration is vulnerable populations, such as low-income Americans and minorities. New customers will be in those socioeconomic groups, and they need more flexibility and the ability to back out of a service if they have a change in income. There's no way for those families to sign up for services that are incredibly expensive and really lock you down."

UNLOCKING

Consumer advocates say unlocking phones—breaking the software locks that tie phones to a single network—is one way to reduce costs for consumers. "Unlocking is a way to let vulnerable populations' dollars go further," says Fazlullah. "It lets people go from provider to provider and decide who's giving them a better deal while being able to retain their phones." Unlocking phones also helps international travelers save money by enabling them to change their SIM cards to those of a local carrier and avoid roaming fees.

People have been unlocking cellphones since the early 2000s, but unlocking has always occupied a legal gray zone due to the Digital Millennium Copyright Act (DMCA), a wide-ranging copyright protection law Congress passed in 1998. Though the DMCA was designed to prevent people from illegally copying and distributing music and movies, the government has interpreted the law's anti-circumvention provisions to also prohibit unlocking. These provisions, which appear in Section 1201 of the DMCA, outlaw people from tampering with software locks that control access to copyrighted works. Since a smartphone's software can be copyrighted, disabling or removing the locks that carriers or phone makers place on that software in order to prevent user access can be considered a breach of the DMCA.

Consumer advocacy groups have long protested this interpretation. They argue that copyright law is being erroneously applied to a situation in which there is no copyright infringement or content piracy. "It's a weird loophole," says digital rights activist Sina Khanifar. "The language used in the DMCA makes the law incredibly encompassing."

Khanifar is a San Francisco–based software programmer and

technology entrepreneur and one of the leading advocates of unlocking. About a decade ago he established a small unlocking business to help pay his college bills, but he stopped in 2005 when Motorola sent him a cease-and-desist order. Khanifar says that at the time he was 20 years old and terrified at the prospect of being punished under the DMCA, which carries penalties of up to $500,000 in fines and five years in jail per offense. Motorola eventually dropped the case after the prominent civil liberties attorney Jennifer Granick—currently the Director of Civil Liberties at Stanford Law School's Center for Internet and Society—gave Khanifar pro bono legal help. Khanifar emerged convinced that companies were leveraging the DMCA to squeeze consumers and entrepreneurs.

After Granick assisted Khanifar, she and other civil liberties activists pressed the Library of Congress, which oversees the U.S. Copyright Office, for an exemption to the DMCA's anticircumvention rules. As head of the Library of Congress, the Librarian of Congress has power to issue DMCA exemptions, but the exemptions expire after three years. Activists wrangled an exemption in 2006 and again in 2009. But in October 2012, Congressional Librarian James Hadley Billington reversed his position, citing the "ready availability of new unlocked phones in the marketplace,"[25] presumably from retailers like Amazon and Best Buy. By January 2013, it was no longer legal to unlock new cellphones.

Consumer advocates are battling that decision through grassroots campaigns. They worry that the ruling sets a dangerous precedent for government intrusion in people's lives and belongings. "The question is, when you buy a phone, do you really own it and its software, and can you modify it?" says Khanifar.

In early 2013, Khanifar started an online petition that requested the White House to legalize unlocking. The petition resonated with "the pretty large percentage of people who have had the experience of trying to change [cellular] networks and not being able to," said Khanifar, and it amassed more than 114,000 signatures in 30 days. In response, the Obama administration issued a supportive letter that said it was "common sense" that "consumers should be able to unlock their cellphones without risking criminal or other penalties."[26]

About six months later, the National Telecommunications and Information Administration (NTIA), a Commerce Department agency that advises the president on telecommunications and information policy issues, formally asked the FCC to require carriers to unlock consumers' devices upon request. That sparked negotiations between the FCC and the Cellular Telecommunications and Internet Association (CTIA), the main trade group for the U.S. wireless industry. At the time, most U.S. carriers already allowed some of their subscribers to unlock their phones, but they hesitated to publicize or liberalize their policies, because unlocking conflicts with their business models. On the postpaid side, unlocking threatens the carrier practice of recouping phone subsidies by signing subscribers to exclusive, multiyear contracts. On the prepaid side, unlocking makes it easier for shady businesses to buy mass quantities of prepaid phones that have been slightly subsidized for marketing purposes, unlock the phones, and sell them for profit. "Consumers and consumer advocacy groups are the only ones who fight for unlocking," says Khanifar. "No big companies ask for it; all the money is on the other side of the table."

In the end, the FCC delivered an ultimatum: "It is now time for the industry to act voluntarily or for the FCC to regulate."[27] In December 2013, the country's five largest carriers—AT&T, Sprint, T-Mobile, U.S. Cellular, and Verizon Wireless—committed to adopting six standards related to unlocking. By March 2014, the carriers had posted their unlocking policies on their websites and started unlocking postpaid and prepaid smartphones and tablets for eligible consumers who requested the service. People on postpaid smartphone plans qualify for carrier unlocking only if they have completed their two-year service contracts, fully paid off their phones on a no-subsidy financing plan, or paid an ETF to get out of their contracts early. Still to come: a system that either notifies subscribers when their devices qualify for unlocking or automatically unlocks the devices remotely; a reduction in the time needed to respond to unlocking requests to two business days; and a process for unlocking devices for military personnel who show proof of overseas deployment.

Khanifar says the CTIA agreement is "definitely a victory" for his campaign, because the new policies and principles will simplify and expedite smartphone unlocking through carriers. But he cautions that it is "only a start," because the reforms don't challenge the carriers' central role in unlocking phones. Khanifar's philosophy is simple: "Once you buy something, you should be able to do with it what you want." By leaving carriers in charge of phone unlocking, the CTIA agreement stops far short of that standard.

Activists are hoping federal legislation will give consumers more control. Until early 2014, Khanifar and his colleagues had backed the Unlocking Consumer Choice and Wireless Competition Act proposed by Senator Patrick Leahy (D-VT) and Representative Bob Goodlatte (R-VA) in March 2013. The Leahy-Goodlatte bill legalizes unlocking only until 2015, when the Librarian of Congress is scheduled to revisit the issue yet again. The tech news site Techdirt said the bill "just punts the issue until later." [28]

But the bill also had two things going for it. It was the most likely to get passed, and it seemed to address a point often overlooked in laws related to unlocking: the need to legalize unlocking as both a business and an individual right. Unlocking requires modifying a phone's software, which consumers can do by downloading and running a special program or by punching in a code purchased online that updates the phone's settings. The first method can be tricky, so for many smartphone users the right to unlock their phones is useless unless they can also hire professionals to do the unlocking for them. But under the DMCA, people who unlock other people's phones face the same penalties Khanifar did in 2005, because they are considered to be violating the law "willfully and for purposes of commercial advantage or private financial gain." [29] Even the Librarian of Congress's exemptions technically permitted individuals to unlock only their own phones.

The threat of a DMCA penalty keeps big companies out of the unlocking business. The shops that do unlock phones operate semi-underground. "It feels seedy to unlock phones," says Parker Higgins, an activist at the digital rights nonprofit the Electronic Frontier Foundation (EFF). "That's unfair for consumers; it should be a very straightforward thing."

Activists thought the Leahy-Goodlatte bill was going to make unlocking straightforward because it allowed consumers to ask "another person" to unlock their phones, meaning users could solicit third-party service provider help, but in February 2014 Goodlatte amended it to exclude unlocking phones "for the purpose of bulk resale." Such a policy was likely intended to thwart the unethical mass purchase and resale of new prepaid phones for profit, but it would also inhibit recyclers and refurbishers from unlocking and selling phones—a limitation that groups such as the EFF say would hurt consumers and the environment by restricting easy access to low-cost, used handsets.

Activists are now promoting another bill. Proposed in May 2013 by Representative Zoe Lofgren (D-CA), the Unlocking Technology Act aims to revise the DMCA so that nothing would be considered a DMCA violation unless it actually infringed copyrighted material. The bill would also add new language to the DMCA that explicitly states that unlocking is not infringement and thus is not a violation of it. "It's really an ideal bill for solving unlocking and other DMCA problems," says Khanifar. The bill is still in the committee review stage.

Activists plan to keep pushing for DMCA reform. Says Khanifar, "These are really important questions that are really central to how we will interact with technology going forward."

PLANNED OBSOLESCENCE

Another type of waste occurs when smartphone makers prod consumers to buy new phones by quickly introducing more sophisticated replacements for current ones, discouraging repairs for those that are broken and malfunctioning, and phasing out software updates.

The iPhone is an oft-cited example of this planned obsolescence. People say Apple essentially replaces the iPhone each year by issuing a new model. Radical updates, such as changes to the phone's screen size, only occur every two years, but each new iPhone offers enough improvements that a number of people ditch their serviceable, one-year-old iPhones every year.

People like iFixit CEO Kyle Wiens also say Apple makes it tricky to fix broken iPhones. iPhones are impressively durable, considering their elegant design, but their glass screens can crack if they are dropped onto hard surfaces. When the glass on an iPhone breaks, the phone's liquid crystal display (LCD) often also needs to be replaced, because the glass and the LCD are glued together. In addition, Apple seals iPhone casings tightly, hindering access to its battery, which can be problematic because all smartphone batteries degrade over time and need to be replaced. Wiens says his iPhone 4S needed a new battery after 11 months, adding that "Apple makes very high-quality products, but it's not interested much beyond the twelve- to twenty-four-month range."

Of course, all smartphone makers prefer that consumers buy new phones instead of repairing their old ones. As the *New York Times* noted in an October 2013 article, the fact that Apple is accused of planned obsolescence more than most other companies is "partly a function of just how big a player it is, and how suspicious consumers become when a luxury product so closely associated with excellence doesn't meet their expectations." [30] Apple has also tweaked the design of newer iPhones (the 5 and later) to make repairs easier and now offers screen replacements for the 5 and 5C as a convenient, while-you-wait service in its stores.

Whether or not the iPhone truly is an example of planned obsolescence, more smartphone makers are launching their phones the same way. HTC, Samsung, and others have established their own flagship smartphone "families" that they update annually with new models. HTC, LG, Motorola, and Samsung also complicate repairs in some of their phones by fusing the LCD to the glass screen or adhering other major components together. A number of new smartphones, including the HTC One (M8), Moto X, and the Nexus 5, lack easily removable batteries. In fact, iFixit considers the HTC One (M8) to be a far harder phone to repair than the iPhone 5C or 5S.

Planned obsolescence is also a software issue, and here Apple performs better than other smartphone makers. Apple typically supports iPhones with iOS updates for three years after launch. John Gruber, the influential Apple commentator, has argued that Apple's comprehensive software upgrade policy proves that the idea that Apple "somehow booby-traps its devices to malfunction after a certain too-brief period to spur upgrades to brand-new products" is a "pernicious lie." [31] In a December 2013 post on his Daring Fireball website, Gruber pointed out, "Where is the mobile operating system that does a better job supporting older hardware than iOS?"[32]

In comparison, Android phone makers typically stop releasing updates 18 months postlaunch. Sometimes hardware limitations prevent software updates. For example, processors on older smartphones may not be robust enough to run newer operating system features. Carriers can also hold up smartphone makers' software rollouts with lengthy approval procedures. But smartphone makers usually avoid updating their old phones for two other reasons: creating, testing, and distributing new software requires time and money, and they want consumers to buy new phones anyway.*

The silver lining of planned obsolescence is rapid innovation. The industry's frenzied pace engenders faster, lighter, and more powerful smartphones every year, or even every few months. Software also gets more intuitive and intelligent with each upgrade. Consumers benefit from these improvements. But despite these perks, people such as Wiens deplore planned obsolescence, because it pushes consumers to spend money on new phones when at least some of them would be content with their current ones. Wiens would prefer that consumers repair their phones rather than replace them. "We're on a freight train of consumerism," says Wiens. "Manufacturers are pushing the pedal to the metal as fast as they can, and that's not necessarily the best thing for consumers, the environment, or society."

ESCALATING E-WASTE

When we rapidly upgrade and discard our smartphones, they contribute to the larger problem of electronic waste—a broad term applied to used electronics that have been thrown away, donated, or sold to a recycler. E-waste is one of the fastest-growing types of municipal solid waste, meaning waste created by and collected from houses and office buildings in towns and cities. A technology research firm, TechNavio,

---

*Google has detached its core apps from the Android operating system so it can keep them current on users' phones (by pushing updates through Google Play) without relying on phone makers or carriers.

calculated that the world generated 58 to 63 million tons of e-waste in 2013. According to the StEP Initiative, a partnership of U.N. organizations, companies, governmental institutions, NGOs, and science organizations focused on fighting e-waste, global e-waste will reach 65.4 million tons a year by 2017.

Cellphones are one of the largest sources of e-waste, because people buy and throw out a disproportionately large number of them. Technological advances, wear and tear, design trends, marketing campaigns, and planned obsolescence make phones obsolete (or seem obsolete) faster than other gadgets. On average, Americans keep their cellphones for just 18 to 20 months. According to the U.S. Environmental Protection Agency (EPA)'s most recent e-waste report, Americans discarded an estimated 135 million mobile devices—meaning cellphones, smartphones, PDAs, and pagers—in 2010. That was more than four times the number of computers and more than five times the number of TVs Americans discarded that year.

Cellphones not only increase e-waste volume, they—like many e-waste materials—can be health hazards. Cellphone circuit boards and batteries usually contain at least one toxic metal, such as lead, which can cause neurological and developmental disorders; cadmium, which can result in kidney, bone, and lung disease; and beryllium, which is associated with lung and skin diseases. These toxic metals are safely contained when a phone is functioning but can be released when they are pulverized, incinerated, or buried in landfills.

Tracking e-waste has inherent challenges, and the lack of reliable data produces widely varying statistics. Estimates of the percentage of used American cellphones that are collected for recycling range from 60 percent (from a 2013 study by the Massachusetts Institute of Technology Materials Systems Laboratory and the National Center for Electronics Recycling)[33] to just 11 percent (the EPA). Once collected, a portion is sent overseas, but here too estimates run the gamut, with environmental advocates saying 50 percent of the e-waste collected by recyclers gets exported and the MIT/NCER study saying the amount is around 10 percent.

These various groups do, however, agree that at least some e-waste is exported. Like conflict minerals, it traverses the globe through a complex chain of people and companies, each propelled by its own motives. A typical chain might start with a dishonest e-waste collector pretending to hold a recycling fundraiser for charity. After people donate their used gadgets, the collector sells them to a broker. The broker then picks them up and ships them out of a nearby port—probably to Hong Kong. From Hong Kong, the cargo will probably get smuggled into China, to small companies that pay the brokers for the scrap.

This chain of events ends in places like the southeastern Chinese town of Guiyu, which is often called the world's e-waste capital. China is a cost-effective place for Americans to send e-waste, because imported goods arrive constantly from China, and the enormous containers that carry them return relatively empty. "It doesn't cost much to ship a load of stuff from New Jersey or Oakland, California, back to China," says Michael Zhao, a video-journalist who made the documentary *E-Waste: Afterlife*, on e-waste in China, in 2011.

Guiyu's proximity to Hong Kong and numerous Chinese factories make it a convenient spot to drop off international e-waste and resell it domestically. In fact, Guiyu and Foxconn City are located in the same Chinese province (Guangdong), about 180 miles apart from each other, and like Foxconn Guiyu relies on migrant labor from poor, rural provinces. "Probably, someone from their village said, 'Hey, come here, it's way better than tending a farm at home,'" says Zhao, who has visited Guiyu several times. Migrants may choose Guiyu over factory jobs, because the town's system of small workshops, each of which typically employs fewer than ten people, allows families to live and work together.

Guiyu is home to 150,000 people who run and staff more than 300 companies and 3,000 individual workshops engaged in e-waste recycling. The town's main trade is immediately obvious to any visitor. "On seemingly every street, laborers sit on the pavement outside workshops ripping out the guts of household appliances with hammers and drills," said one CNN report. "The roads . . . are lined with bundles of plastic, wires, cables, and other garbage."[34]

Guiyu laborers harvest these parts from computers, computer peripherals such as monitors and printers, TVs, and cellphones. Zhao says workers follow the same basic steps to disassemble both computers and phones. They remove the outer metal or plastic casing to uncover the circuit board and other components, then cook the circuit board to melt its solder. The cooking process produces toxic fumes, but it allows the chips to be removed from the board fully intact. Those chips are dusted off and resold, often under the pretense that they are new.

Worker dismantling toner cartridges, covered with toner, in Guiyu, China (*Basel Action Network/Flickr*)

When Zhao first visited Guiyu in 2006, he saw another set of workers immersing those circuit boards in buckets of burning acid to extract valuable metals, such as gold, silver, and copper. The buckets emitted thick, toxic smoke. Zhao's film shows one of these workers warning onlookers, "Go away, you can't handle this; it's too choking. The sky is burning black." [35] Zhao recalls: "I tried to take a quick look at what was going on inside one of the vats, but the smoke burned my eyes immediately. . . . Those guys wear a pair of gloves [to protect their hands from the acid] and that's it. I don't know how they can deal with it for so long."

Zhao says these acid vat plants paid more than Guiyu's other entry-level e-waste jobs and thus were able to find willing workers. "These people are mostly concerned with economics," explains Zhao. "They're not thinking how bad their health will be ten years from now."

Naturally there are health consequences to both living and working in Guiyu. The town has been processing e-waste since the early 1990s, and years of burning, melting, and acid-stripping e-waste have left the air, water, and soil polluted with lead and other toxic metals. "You smell Guiyu before you get there," says Zhao. "It smells of burning wires and circuitry." A 2010 study conducted by the nearby Shantou University Medical College found as many as eight out of ten Guiyu children had lead poisoning, which can affect mental and physical development.

The Chinese government banned the import of toxic e-waste back in 2000, but Guiyu continues to receive and process e-waste due to corruption. The local government reaps a "significant portion . . . of its annual revenues" from the practice and does not enforce regulation, according to a 2013 StEP Initiative report.[36] "Beijing wants to get rid of [the illegal e-waste business], but the local government isn't ready yet, because of economic considerations," says Zhao. "If [local government officials] follow the rules, it's basically death to the local economy, and they haven't found another business to replace it."

There has been progress. In 2013, China launched a ten-month campaign, Operation Green Fence, to firm up enforcement of its e-waste laws. The initiative stationed customs officials at Chinese ports to check the quality of foreign scrap. Shipments that exceeded the Chinese legal limit of 1.5 percent of contaminants were returned to their originating countries, causing a significant drop in e-waste imports. On Zhao's most recent Guiyu trip, in 2013, he saw fewer dilapidated workshops and no acid vat plants. He says, "There are still piles of junk [on the ground], but they are much smaller and less obvious than a few years ago."

Another change in recent years is that much of the e-waste in Guiyu now comes from within China. China's burgeoning middle class is buying more gadgets, and the country is now home to at least 1.2 billion cellphones, many of which will be discarded in coming years. "People in cities often have two or even three cellphones at the same time, and they upgrade fast," says Zhao. "China has become a major source of e-waste by itself, and the majority of these hundreds of millions of cellphones will stay in China." The influx of domestic e-waste hinders Guiyu from implementing more sweeping environmental reforms.

"The whole place is still built on this dirty process, and it's not going to get really clean anytime soon," says Zhao. "E-waste is an important issue, but there are so many other [health and safety] issues in China."

Guiyu is one of China's most internationally famous e-waste sites, but informal e-waste markets and slums exist elsewhere in China, and there are large e-waste dumps in or near Bangalore, Chennai, and New Delhi in India, Karachi in Pakistan, and Lagos in Nigeria. Agbog-bloshie, a suburb in the Ghanaian city of Accra, is sometimes called the e-waste capital of Africa. Agbogbloshie originally took in used electronics, mostly from Europe, for resale in its local market. Similar to China's turn inward, Agbogbloshie is busy these days processing Africa's proliferating domestic e-waste. In 2010, residents of Benin, Côte d'Ivoire, Ghana, Liberia, and Nigeria generated between 50 percent and 85 percent of the e-waste in their countries, according to a 2011 report.[37] As in Guiyu, used-electronics imports from industrialized countries exacerbate West Africa's local e-waste problem. The report notes that up to 30 percent of West Africa e-waste originates from residents using "used [electronics] of unclear quality"[38] imported primarily from Europe, but also from Asia and North America.

Looking for metals to reclaim in Agbogbloshie, Ghana (*Fairphone*)

E-Waste Legislation

International treaties exist that forbid the trade of nonfunctioning e-waste. However, the United States is not a party to them. The Basel Convention is a U.N.-sponsored treaty that regulates the export and import of hazardous waste. It became effective in 1992 and has been ratified by 181 countries that pledge not to export hazardous waste unless the receiving country gives prior written consent and the trade is done in an "environmentally sound"[39] manner. The United States signed the Basel Convention in 1990, but never ratified or implemented the treaty, so it is not bound by its terms.

In 1994, when it became clear that the Basel Convention wasn't enough to stop global e-waste dumping, the countries that ratified the treaty introduced an amendment known as the Ban Amendment. The Ban Amendment specifically prohibits the export of hazardous waste from wealthy countries that are part of the Organization for Economic Cooperation and Development (OECD) to non-OECD countries. So far, 78 countries, including the European Union, have ratified the amendment. It will be entered into force if thirteen more countries that were party to the Basel Convention in 1994 sign it.

The United States has not signed or ratified the Ban Amendment. Environmental activists say corporate interests prevent the government from taking action. "There tends to be a kneejerk response that if something restricts exports, it's antibusiness and harms our economy," says Barbara Kyle, the national coordinator of the Electronics TakeBack Coalition (ETBC), a San Francisco–based coalition of environmental and consumer organizations that promotes responsible recycling of consumer electronics. E-waste may still leak from European countries to places such as Agbogbloshie despite the Ban Amendment, but as Kyle says, "They're definitely [doing] better than us [at tackling e-waste exports]."

Currently the United States has only one e-waste export rule, and it relates to cathode ray tubes, the glass vacuum tubes that were used in TV sets and computer monitors before flat screens became popular. Cathode ray tubes each contain several pounds of lead, so companies are supposed to notify the EPA and get permission from the receiving country before exporting them for recycling. In July 2013, a federal court fined a Colorado-based company called Executive Recycling $4.5 million and sentenced its CEO to two and a half years in prison for smuggling more than 100,000 used cathode ray tubes to China and other countries. The ruling, which was the first U.S. conviction for an e-waste exportation crime, scared dishonest recyclers, but experts say most e-waste infractions continue to go unnoticed and unpunished. "Electronics recycling is almost entirely unregulated as an industry [in the United States]," says Kyle.

Environmental advocates hope a bill called the Responsible Electronics Recycling Act (RERA) will bolster U.S. e-waste export regulations. RERA aims to create a new category of e-waste called "restricted electronic waste." Gadgets that contain toxic materials would be considered restricted electronic waste and banned from export to non-OECD or non-EU nations. Violators would incur criminal penalties. Kyle says the bill was designed to be consistent with the Basel Convention and the Ban Amendment. "Similar aims would be accomplished, just through legislation and not a treaty," she says.

RERA was first introduced in Congress in 2011, but the bill never moved out of the committee stage, despite bipartisan support. In July 2013, RERA was reintroduced with the backing of a large coalition of electronics recyclers that have U.S. operations. The group, which is called the Coalition for American Electronics Recycling, has more than 125 members, including the world's biggest electronic recycler, Britain's Sims Recycling Solutions, and North America's biggest recycler, the Houston-based Waste Management. The coalition is championing RERA as a job creator. It estimates RERA will foster up to 42,000 jobs in the United States by mandating that more e-waste be recycled domestically. Apple and Samsung are also listed on the bill as official supporters.

Activists are confident RERA will pass now that it has strong business backing. But another big industry association, the Institute of Scrap Recycling Industries (ISRI), opposes the bill and has lobbied against it in the past. In July 2013, ISRI criticized RERA for "identifying environmental risk based simply on geographic location rather than responsible operating practices," an approach it called "outdated."[40] ISRI also said the bill violates U.S. trade obligations by discriminating against non-OECD countries. But Kyle says some rules are necessary to prevent the worst forms of e-waste from being sent to countries that have poor records of managing toxic materials. "RERA does put restrictions on what form of stuff can get exported, but it is not anti-export," she insists.

SMARTPHONE TRADE-INS

One way to counteract e-waste is to reuse gadgets and prevent them from becoming e-waste in the first place. Smartphones are uniquely qualified for reuse, since they are more likely than other gadgets to be refurbished and resold due to consumer demand. A smartphone can have several lives and owners without requiring major changes to its hardware or software. Usually it just needs to be reset to its default software to clear personal data and to be outfitted with a new SIM card, so it can operate with a different phone number. In terms of environmental impact, reuse is preferable to recycling, because it is a more efficient use of resources.

The challenge with smartphone reuse is that people tend to hang on to their old phones. People either don't realize that the phones have monetary value, are concerned about erasing personal information from them, or want to keep them as backup devices in case their new phones break or malfunction. Kyle calls this the "dead cellphone drawer" problem. That's the drawer where people store their most recent phone, or maybe even every cellphone they've ever owned.

A number of companies are trying to pry open these drawers by offering consumers money to trade in their old smartphones. Many of these companies operate online. They solicit used smartphones via their websites, ask consumers to mail in their phones, and send payment via mail or an online service such as PayPal.

Gazelle is the largest of these online trade-in sites. The Boston-based company buys, resells, and recycles used smartphones, tablets, iPods, and Mac and Macbook computers, with smartphones making up the majority of its transactions. Gazelle's business illustrates why smartphones have strong reuse potential. For logistical reasons,

Gazelle purchases devices only from consumers located in the United States, but the smartphones it buys go on to have second lives all over the world. After wiping the phones' software to remove personal data, Gazelle lists phones that are in demand and in good condition on eBay and Amazon. Gazelle CEO Israel Ganot says 30 percent of Gazelle's inventory is resold this way. The remaining 70 percent goes to around 30 wholesalers for resale, mostly in Africa, China, India, and Latin America.

Consumers in developing economies are eager buyers of used smartphones. Carriers in those regions usually don't subsidize phones, so new smartphones can cost as much as $800 to $1,000 when import taxes or other tariffs are added. These consumers may also have limited access to smartphones because of regional supply shortages or distribution obstacles. "If you live in the middle of China, there are no Apple stores," notes Ganot.

Ganot says he can always find buyers for Gazelle's smartphones. What he needs are more phones to sell to them. The smartphone subsidy system is one factor preventing more Americans from trading in their phones. Ganot says Americans don't recognize how valuable their smartphones are because we pay relatively little for them. "Most of us, when we buy a subsidized phone from a carrier, will pay $200," he explains. "So, consumers think [their phones] are worth nothing after using them for a few years, but actually they can get as much as what they paid originally [if they trade them in]."

Companies like Gazelle were early to spot smartphones' resale value. Most carriers and phone makers took longer to launch trade-in programs that target resalable devices and offer people a financial reward. Initially smartphone companies offered only takeback programs that recycled people's old, beat-up phones (and offered no compensation). Some takeback initiatives benefited charities, such as Verizon's HopeLine program, which has donated used phones to organizations against domestic violence since 2001. Others were state-mandated. California, Illinois, Maine, and New York have recycling acts that require companies selling cellphones in their states to take back used phones and recycle them. In other states and countries, smartphone companies operate takeback initiatives on a voluntary basis.

Today almost all of the major smartphone companies offer trade-in programs, usually managed by outside logistics providers. Trade-in and takeback programs are the main ways smartphone companies burnish their environmental credentials and counteract the e-waste they help create. Publications such as *Newsweek* regularly cite Sprint as a "green" company, due to its extensive trade-in and takeback efforts, which Sprint says have resulted in the collection of more than 50 million wireless devices. Sprint's site tells visitors they can "make a difference today" by trading in their phones and keeping them out of landfills.[41]

Sprint isn't just interested in being green. Trade-in programs are another way for carriers to expand their margins—in this case, by affordably acquiring phones that can be refurbished and resold to consumers, either in the United States or abroad or deployed as replacement handsets in carriers' phone-insurance programs. Carriers make more profit from these transactions than they do selling new subsidized smartphones, because there is no subsidy involved in a used-phone trade-in. As the market researcher Pyramid Research published in a recent report, "The increasing weight of smartphone subsidies on mobile operators' profitability and the growing demand for used smartphones in emerging markets create fertile land for buyback and trade-in programs."[42]

Like the carriers, smartphone makers buy back their old phones for several reasons. They too want to encourage phone upgrades by giving consumers funds to buy new phones. Phone makers may also want to resell these used phones in developing economies, harvest the phones for replacement parts for repairs, or just keep as many phones as possible off the secondhand market, so consumers will have to buy new phones.

Certified Recyclers

Trade-in programs might seem ideal solutions to the smartphone e-waste problem but collecting phones is only half the battle. If trade-in programs don't repair, resell, and recycle the phones they receive responsibly, they will intensify the e-waste problem rather than help alleviate it. As Kyle puts it, "There are two very different [e-waste] issues: how hard a company tries to collect stuff, and then what it does with it." The latter question is particularly pertinent for recycling smartphones. Takeback programs and charitable recycling drives are likely to net smartphones in poor condition, and even trade-in programs can attract plenty of junk. Before Gazelle streamlined its business in 2012, it accepted 22 different types of gadgets and had to send 10 percent of its inventory to recycling, according to Ganot.

Fortunately, even broken, unusable smartphones have considerable value. They are more lucrative for recyclers than other forms of e-waste because they have a relatively large amount of precious metals for their small size. Their circuit boards may contain gold, palladium, platinum, and silver, as well as copper and nickel, which are not precious metals but do have monetary value. The EPA has calculated that recycling a million cellphones can recover 35,274 pounds of copper, 772 pounds of silver, 75 pounds of gold, and 33 pounds of palladium. Recycling companies will pay more than $15 per pound for smartphone circuit boards so they can reclaim these metals, says Joel Urano, president and CEO of Capstone Wireless, a Dallas-based cellphone repair, refurbishing, and resale company.

Environmentalists know recyclers are motivated to collect smartphones, but they worry that unscrupulous recyclers will cherry-pick valuable parts, such as circuit boards, out of their scrap and dump the rest in the trash or export it to places like Guiyu and Agbogbloshie. Activists say consumers in search of ethical recyclers should look for certification credentials, for instance, from Responsible Recycling (R2) or e-Stewards.

Of the two, e-Stewards has stricter ethical and moral requirements. Both programs oblige companies to maximize the reuse and resale of the e-scrap material they receive, to safeguard their workers' health and safety, and to wipe data from phones they receive to protect consumers' privacy and security. But e-Stewards is the only standard that prohibits the export of hazardous e-waste to developing countries. R2-certified companies can send used electronics internationally as long as the receiving country legally accepts such shipments and the company "demonstrate[s] compliance of each shipment with the applicable export and import laws."[43]

The two standards also diverge on the matter of using prison labor to process e-waste—a practice that appeals to some recyclers because of its low cost. To employ prison labor, recyclers hire UNICOR, a government-owned corporation established in the 1930s to give inmates job training. UNICOR started using prisoners to recycle computers and other gadgets in Florida in 1994 and expanded its operations to prisons in several other states in 1997. It says it teaches inmates marketable skills, but early tasks included smashing TVs and computer monitors with hammers—and without protective equipment. By the early 2000s, workers were concerned about respiratory problems and rashes. Prompted by lobbying on their behalf, the Justice Department's Office of the Inspector General investigated UNICOR recycling facilities across the country. The resulting report, published in 2010, said UNICOR exposed workers to 31 metals, including cadmium and lead, and "concealed warnings"[44] about those hazards. UNICOR started making improvements in 2003 and by 2009 its e-waste operations were being operated safely "with limited exceptions."[45] Now UNICOR is R2-certified and R2 companies can hire UNICOR to recycle e-waste, but the e-Stewards standard expressly prohibits the use of prison labor, due to ongoing concerns about health risks for prisoners and data-security risks for consumers and companies.

Smartphone recycling is a multistep process that involves some human labor and a lot of automation. Companies send smartphones to e-Stewards and R2 recycler ECS Refining in huge boxes that can hold 10,000 units. Often these phones are relatively intact but damaged beyond repair. ECS employees inspect the phones and remove their batteries, which must be handled separately, because they can cause fires or small explosions if punctured. After this triage is complete, shredder and grinder machines chop the rest of the phone into two-and-a-half-inch-long pieces. Those are spit out onto a conveyor belt, where magnets isolate glass and plastic from metal, and then separate different types of metals into piles.

The entire process, from triage to the sorting of shredded material, takes only a few hours, according to Jeff Bell, the general manager at ECS's Mesquite, Texas, facility. Smartphones take a little longer to triage than basic phones, because their casings are harder to open and their batteries are often concealed deep within the phone. "The iPhone is a sealed unit," says Bell. "You have to touch smartphones a little more to get the batteries out." Once these flammable items have been removed, Bell says, his plant, which is one of three processing centers ECS operates in Texas and California, can shred 15 to 25 tons of smartphone material per hour. ECS sends the shredded material to refiners in Belgium and Canada that smelt it to recover precious metals. Those companies pay ECS based on the weight of the precious metals they are able to reclaim.

Under e-Stewards rules, ECS can export e-scrap, because it is not sending junk but rather raw materials, which other companies use to create new products. "There are no residuals when we're done [processing a smartphone]," says Bell. "The next people take [what we produce] and turn it into a new form." Recovered aluminum and steel are often sold at metals markets. Plastic smartphone casings have less value, but they can be sent to plastics processors that grind them into pellets, which manufacturers then buy to make new goods. Used smartphone batteries can be resold, too. Bell says various companies will buy batteries that still hold 85 percent to 90 percent of their charge and use them as replacement parts. Dead batteries go to recyclers that specialize in smelting batteries for metals, including lithium and nickel. ECS even buys smartphone chargers and charging cords, because they contain copper and steel, which can be reclaimed after shredding and chopping. Glass is the one smartphone material that is difficult to resell, because it needs to be pristine to retain its value. Says Bell, "We can resell glass if it's usable, but typically we're recycling it [at a glass recycler]."

ECS essentially functions as a one-stop shop for e-waste processing. Bell says approximately 10 companies in the United States can handle the same breadth and volume of material under one roof. The number of e-Stewards recyclers that can do the same is much smaller. The reason is simple: recyclers must invest considerable time and expense to get certified and stay certified. Bell's plant has sophisticated environmental controls, including systems that monitor air quality throughout the facility and collect the dust generated by the grinding machines. ECS also provides hazardous-material respirators and uniforms for its employees and launders their uniforms to ensure they are kept clean. The company even tests its employees' blood to certify that they are not being exposed to high lead levels from the circuit board grinding process.

There's also the matter of record keeping: e-Stewards regulations emphasize transparency and accountability in business dealings. ECS has to track the materials it buys and sells, through inventories and shipping manifests, and share those records with its e-Stewards customers. All e-Stewards companies pay for a third-party auditing firm to check their compliance with the program's rules and they also send Basel Action Network (BAN), the nonprofit that oversees e-Stewards, an annual licensing and marketing fee that ranges from $500 to $90,000, depending on their revenues. Noncertified recyclers have lower operational expenses and can afford to pay more for used materials. That gives companies that don't care about rigorous recycling ethics a big incentive not to select a certified recycler. Capstone CEO Urano says prospective clients have told him, " 'These [other] guys are paying more and will recycle everything.' . . . Those companies don't ask [those recyclers] if they're R2 or e-Stewards certified."

A few large retailers, such as Best Buy and Staples, use certified recyclers for their takeback and trade-in programs. Smartphone companies have varied policies. Apple utilizes Sims Recycling Solutions, which is e-Stewards and R2 certified, for its iPhone recycling program. Samsung and LG have been e-Stewards "enterprise" members since 2010 and 2011, respectively, which means they've agreed to use e-Stewards recyclers when possible for their recycling efforts. In 2013, LG upped its commitment and now sends all of its U.S. e-waste to e-Stewards recyclers, leading BAN to call it "the most responsible electronics company operating in the United States when it comes to managing e-waste."[46] Sprint and Verizon use eRecycling Corps, an R2-certified recycler, for their device buyback and trade-in programs and T-Mobile says its recyclers are either R2- or e-Stewards-certified.

Other smartphone companies say they responsibly recycle any e-waste they acquire, but they don't publicize that they use certified recyclers, so it's unclear how rigorous their standards are.

E-waste Solutions

Smartphones will eventually follow the same trajectory as basic cellphones. As more people upgrade to smartphones around the world, more smartphones will be discarded and/or recycled, and fewer of them will be successfully resold and reused. Bell says the number of first- and second-generation iPhones and Galaxy S phones arriving for recycling at his ECS facility has been increasing since 2013. Trade-in programs will amplify the need for ethical recycling systems in developing economies, because once a used smartphone has had a second or third life in a developing economy, that country needs to be able to properly dispose of the phone.

Mitigating this e-waste entails changes in design, manufacturing, recycling, and policy. Governments could mandate the types of materials used in smartphones. The European Union has a "restriction of the use of certain hazardous substances" (RoHS) directive that took effect in 2006 and has been implemented by EU member states. RoHS requires manufacturers to limit the use of lead, mercury, cadmium, and chemicals called brominated flame-retardants in a range of gadgets, including cellphones and smartphones. Apple, LG, Motorola, Nokia, Samsung, and Sony have complied by phasing out brominated flame-retardant plastic, which can cause cancer, liver damage, and thyroid dysfunctions, from their more recent phones. These phone makers have also removed a few other high-risk chemicals, such as polyvinyl chloride (PVC), which can cause reproductive, developmental, and other health problems. More countries could enforce similar regulations.

Another e-waste mitigation strategy could be called the iFixit argument. This approach requires phone makers to design smartphones so they can be easily opened and repaired in case of problems. They could also be designed in ways that would streamline the dismantling process at the end of their lives. Smartphones are extremely complex devices, but if phone makers were able to simplify disassembly and component removal, workers in places like Guiyu and Agbogbloshie wouldn't have to resort to crude methods such as burning in order to process e-waste.

Perhaps the most intriguing idea is to increase producer responsibility for smartphone e-waste. Laws could order smartphone makers to collect and properly dispose of the phones they produce. Since 2003, the EU has had a Waste Electrical and Electronic Equipment (WEEE) directive that demands phone makers take back used phones sold in the EU. Loopholes have reduced its efficacy, but the European Union has been working to close them. South Korea has had a similar law since 2005, and India passed one that went into effect in 2012. Cell-phone recycling laws in California, Illinois, Maine, and New York also require retailers and manufacturers to take back their used phones. Other states either don't have takeback laws or have laws that target larger gadgets, such as computers and TVs.

Most of the major smartphone makers do accept and recycle used phones, either by law or voluntarily, but only in certain regions. Take-back programs are typically weaker in non-OECD countries. Environmental advocates say "extended producer responsibility" laws would give phone makers financial incentives to make their designs more environmentally friendly. Such laws could enable companies that build phones with less toxic materials to save money on their recycling costs, for example. "If manufacturers put good materials into their devices, they should be able to realize a profit," says Kyle. "The economics of recycling should be part of design."

Consumers have to do their part, too. As Gazelle's Ganot points out, "The customer is the one who ultimately decides to throw his smartphone in the trash or put it in a drawer." Consumers who want to reduce e-waste and wireless waste also need to curb their spending. Myriad forces and companies push us to supersize our smartphone plans and upgrade our smartphones, but in the end we make the decision to buy.

6

Health

The nationwide cellphone cancer scare began in January 1993 when David Reynard went on *Larry King Live*. He had recently lost his wife, Susan, to a brain tumor, which he believed had been either caused or accelerated by her frequent cellphone usage. Before her death Susan had filed a lawsuit against her phone's manufacturer, Japan's NEC, carrier GTE (later part of Verizon), and the local retailer that sold the phone. It was the first time a person had sought compensation for brain cancer caused by cellphone radiation.

Though a Florida judge later dismissed the lawsuit, citing insufficient scientific evidence, Reynard's appearance on King's widely watched show rattled the country. Frightened by the cancer connection, some users canceled their cellphone service. Consumers who had planned to buy cellphones delayed their purchases. Investors quickly sold off their shares of mobile communications companies.

The CTIA, the U.S. wireless trade group, said the panic was a case of emotion trumping science. Cellphones use radiofrequency (RF) energy, a form of electromagnetic radiation, to send and receive voice and data signals to cell towers, and the mere mention of the word *radiation* frightened many people. But the RF energy cellphones emit is too weak to cause the type of DNA damage that provokes cellular mutations and can lead to cancer.

Still, the question lingers: Could repeated exposure to RF energy indirectly cause health problems? In pursuit of an answer, health organizations, governments, academics, and cellphone companies have spent tens of millions of dollars on hundreds of studies. Today there is still no conclusive scientific evidence that cellphones cause cancer, but there have been enough mixed signals that many people, including some prominent scientists and doctors, have not ruled out a cellphone–

cancer link, either. "More studies involving humans are still needed," says Dr. Henry Lai, a professor emeritus of bioengineering at the University of Washington and a noted cellphone radiation researcher.

The most ambitious study yet of the relationship between cellphone use and brain tumor risk yielded contradictory findings. Known as the Interphone Study, it compared the cellphone histories of two groups: a case group of people aged 30 to 59 who had brain tumors and a control group of tumor-free people who lived in the same region and were the same gender and roughly the same ages as the people in the case group. The idea was to see if past cellphone use could explain why the case group had developed brain tumors while the control group had not.

Interphone kicked off in 2000. The scientific and medical communities had high expectations for it, since it was a project of the International Agency for Research on Cancer (IARC), an arm of the World Health Organization (WHO), which is globally recognized as the foremost authority on identifying cancer causes. The IARC tapped a consortium of researchers from 13 countries, including Australia, Canada, Japan, and eight European nations (the United States was not involved) to lead the project, which involved approximately 10,000 participants over four years and ultimately cost more than $30 million. Yet Interphone's conclusions were inconclusive, and the researchers fought for years over the best way to present their results.

These disputes were evident in Interphone's report, which was belatedly published in 2010. It said that participants showed no overall increased risk of brain tumors from cellphone use, but also that there were "suggestions" that the heaviest 10 percent of users—people who had used their phones at least 30 minutes a day for 10 years or more—had a 40 percent higher risk of glioma, a type of brain tumor.[1] The Interphone report ultimately downplayed this finding, noting that potential "biases and errors . . . prevent a causal interpretation."[2] But media coverage of the study focused on the glioma statistic. Newspapers around the world ran disquieting headlines such as SHOCK FINDING: CELL PHONE-CANCER LINK (*New York Post*)[3] and WATCH THAT THING BY YOUR HEAD (*The Globe and Mail*).[4]

People grew even more wary of cellphones the following year, when the IARC classified RF energy as a possible carcinogen. To formulate its decision, the agency had convened an international panel of experts to review dozens of published studies. The experts also drew heavily on the Interphone glioma data. The IARC's ruling basically stated that cellphones could possibly carry some cancer risk, which placed them in the same broad IARC category as coffee and pickles. Nonetheless, the IARC's classification reignited the global debate about cellphone safety. The *New York Times* ran an editorial that read, "Cell phone users have every right to be befuddled."[5] The *Wall Street Journal* published a much harsher editorial that chastised the IARC for promoting a "needless cancer scare" and for "using its public health platform to exaggerate minuscule risks."[6]

Just a few months after people had digested the IARC news, two European studies eased fears. The first one, conducted between 2004 and 2008 by the Swiss Tropical and Public Health Institute, focused on approximately 1,000 7- to-19-year-olds living in Denmark, Norway, Sweden, and Switzerland. It compared the cellphone usage of children and teens who had been diagnosed with brain cancer with the cellphone habits of their healthy peers. The second study was sponsored by the Danish government and analyzed the health and phone records from 1990 to 2007 of more than 350,000 adult Danes. It compared brain cancer rates between cellphone subscribers and non-subscribers.

Neither study found a cause-and-effect relationship between cellphone use and brain tumor risk, but both studies attracted considerable attention. The Swiss study was seen as important because children's skulls are thinner and have more water content in their bone marrow than adults' skulls. That could theoretically make children more vulnerable to harmful RF exposure, since phone radiation could penetrate deeper into their brains. The Danish study was deemed significant because of its size, relatively long study period, and unique methodology.

Doctors and scientists have criticized some cellphone health studies, including Interphone, for relying on people's ability to remember their cellphone activity from long ago. Experts warn that "recall bias" can skew a study's results. The Danish study avoided this problem. Instead of asking people whether they owned cellphones and when they acquired them, researchers pulled phone subscription data—though not usage data—directly from carriers' billing records. But that study had its own limitations. Because the Danish researchers didn't interview their study subjects, they knew only whether a person had a cellphone plan; they could not discern how much the person actually used the phone, which is valuable information in any cellphone health study.

There has been a dearth of meaningful research on cellphones and cancer since 2011. Two ongoing studies appear promising because of their size, focus, and methodology. MOBI-KIDS is a companion study to Interphone, with some of the same researchers. MOBI-KIDS is similar to the Swiss study, though larger in scale; researchers have said they plan to recruit as many as 3,000 young people aged 10 to 24 in 16 countries, mostly in Asia and Europe and question them about their past cellphone exposure. About one third of the participants have brain tumors, and the remaining two thirds are healthy. MOBI-KIDS was launched in 2009, and its results are expected to be available in 2015 or 2016.

COSMOS aims to be the most comprehensive cellphone health effect study ever. It is similar to the Danish study but larger in scope. COSMOS plans to follow its participants (aged 18 to 69) for 20 to 30 years. The study, which began in 2010, is monitoring about 300,000 people in five countries: Britain, Denmark, Finland, the Netherlands, and Sweden. Like the Danish study, COSMOS is evaluating its subjects' behavior in real time, so there should be no recall bias. In contrast to the Danish study, COSMOS consists of volunteers, so it will be able to track more details about its participants' phone use through questionnaires, phone bills, and health records. COSMOS also addresses perhaps the most common criticism of these kinds of studies—their brevity. Since brain cancer can take as long as four decades to develop, studies can't be considered authoritative unless they gather information for decades, too.

One caveat with long-term phone health studies is that they must release results on a rolling basis or their conclusions will be outdated. Phone technology advances so quickly that study findings can be irrelevant if they're more than a few years old. Devra Davis, an outspoken epidemiologist who has made cellphone health risks her leading cause, says the industry's rapid pace hinders analysis. As she describes in her 2010 book, *Disconnect: The Truth About Cell Phone Radiation*: "Because cellphone use has grown so fast and technologies change every year, it is as if we are trying to study the car in which we are driving."[7]

SAR LEVELS

No health studies have specifically assessed whether smartphone users absorb more or less RF energy overall than those who use basic phones. Lai says newer generations of phones tend to emit less RF energy than older ones, because they are more power-efficient—an encouraging fact for smartphone users. But he also says that the lack of formal studies on the subject makes it difficult to conclude that 3G is "more potent" than 4G. In addition, the frequency bands a network uses, which usually vary from carrier to carrier and from region to region—for example, between the United States and Europe—can also affect the amount of RF energy a phone emits.

When assessing possible RF energy health implications, the most important piece of information is how much emitted energy a user absorbs. Lai says body absorption depends on a number of factors, including the user's distance from the nearest cellular network base station, since phones emit more RF energy when they have to transmit signals across long distances. Other important factors are the duration of use and the phone's proximity to the user. On a smartphone, those two factors are likely to temper each other. Smartphone users tend to interact with their phones constantly, but since they're usually doing things like checking e-mail, browsing websites, and playing with apps, they keep their phones far from their heads. Even people who take a cautionary stance on cellphone radiation don't consider that type of usage particularly risky. "RF energy decreases exponentially with distance from the body," explains Lai. Holding a phone two centimeters away from your ear instead of one would cut RF exposure by 75 percent. A person who primarily uses his smartphone for nonvoice applications would still absorb RF energy in his hand(s), but sensitive areas, such as the person's torso and head, would absorb little.

Consumers don't need to try tracking all these data points themselves. Smartphone makers calculate an RF exposure number for each of their phone models and disclose that information as a value called "specific absorption rate." SAR is a measurement of the RF energy that people absorb when they hold actively operating cellphones close to their bodies, and government regulators in each country specify a maximum SAR value for user safety. In the United States the FCC has this responsibility. Its SAR guideline for operation near a user's head is 1.6 watts (of RF energy) per kilogram or 2.2 pounds (of a person's body weight), which means a cellphone must have a SAR under that level in order to receive FCC certification and be sold legally in the United States. The iPhone 5S has a SAR of 1.18, the iPhone 5C's is 1.19, and the iPhone 5's is 1.25 on AT&T. Samsung's Galaxy S4 has a SAR of 0.75 to 0.98, depending on the carrier model.

Whether consumers should pay attention to a phone's SAR and how phone makers, carriers, and retailers should convey SAR information are subjects of considerable debate. Lai says, "common sense suggests" that less RF exposure—as indicated by a lower SAR—is better for consumers' health.

Phone makers list SAR values in the user safety manuals they package with new phones. They also file this data with the FCC, which in turn makes SAR information available to the public on its website. But people need to know a phone's FCC ID in order to look up its SAR, and FCC IDs aren't easily accessible; they are typically stored inside the phone's battery compartment and/or printed inside the user manual. Consumer advocates say carriers and retailers should clearly display SAR information in stores. They want people to be able to take the value into account when they are shopping. Lai, too, believes providing cellphone/smartphone SAR information to consumers is useful. He says that it is "not a perfect gauge," but because it "at least gives a general idea of [a user's RF] exposure," and is used industrywide, it is the best index of RF energy absorption that currently exists.

The FCC, carriers, and phone makers oppose the idea of prominent SAR labels or signs, saying they represent the maximum amount of radiation phones emit in worst-case scenarios rather than the RF energy a phone generates during normal use. In other words, a higher

SAR number doesn't necessarily mean a greater radiation risk, so people shouldn't use the SAR level to compare cellphones.

This same basic argument springs up whenever anyone introduces a phone radiation-related consumer rights law. In 2010, California, Maine, and San Francisco separately proposed legislation that would have listed a phone's SAR value or a radiation warning either on the phone's packaging or on store displays. CTIA lobbied against all of the bills, and the California and Maine bills never passed. San Francisco's ordinance did but was never implemented, because CTIA successfully sued to block its rollout. CTIA filed a lawsuit against San Francisco, saying the ordinance would force retailers to make misleading statements about cellphone safety, which would violate their First Amendment right to free speech. San Francisco lawmakers accused the wireless industry of "trivializing the First Amendment" and "quashing the debate about the health effects of cellphone radiation." [8] The fight dragged on for nearly three years, until San Francisco agreed not to enforce the ordinance and CTIA agreed not to ask the city to pay its attorneys' fees.

In 2011, Oregon and Pennsylvania introduced their own cellphone radiation bills, and in 2012 Representative Dennis Kucinich (D-OH) backed a more extensive federal bill. All three initiatives were ultimately shelved. In 2014, there was another burst of state-level activity in Hawaii and in Maine, which tried, but failed to pass a modified version of the bill that was rejected in 2010.

Some consumer advocates have since shifted their focus from public information campaigns to pushing the FCC to update its cellphone radiation standards. The FCC established its radiation testing rules in 1996 and has not revised them since. These tests usually entail pouring liquid into a plastic mold shaped like a human head and body and torso. Technicians place smartphones near the dummy head and mechanical probes measure how deeply cellphone signals permeate the liquid, which simulates the electrical properties of human tissue. Those measurements are used to calculate a phone's SAR.

Consumer advocates say the testing methodology has two major flaws: the model for the head is a (theoretical) large, adult male who stands six feet two inches tall and weighs more than 200 pounds. There is no testing model for young people. As Davis, the epidemiologist and activist, wrote in her book, the FCC used "a big-guy brain and body as the basis for all standards for everyone."[9] The second problem: by using liquid as a stand-in for the human brain, cellphone radiation tests erroneously assume people's brains are uniform in consistency.

Lawmakers have also expressed concern about the adequacy of these tests. In 2011, several members of Congress asked the U.S. Government Accountability Office (GAO), a congressional unit that audits federal programs, to evaluate cellphone health hazards and the FCC's radiation safety standards. Following a yearlong inquiry, the GAO issued a report that recommended the FCC "formally reassess and, if appropriate, change" its cellphone testing requirements to ensure phones are meeting SAR guidelines "in all possible usage conditions."[10]

According to the GAO, one red flag was the way people hold active phones against their bodies, such as when people conduct conversations through Bluetooth headsets while leaving their phones in their pants or shirt pockets. Since the FCC does not ask phone makers to test this type of usage, instead measuring "on the body" RF exposure using smartphone belt clips or holsters, people could be exposing themselves to RF energy above the FCC limit, the GAO said. People are supposed to keep active phones a specified distance—usually a centimeter or two—away from their bodies, but many people don't realize this. Phone makers bury these specifications in their user safety manuals, which consumers typically discard without reading.

The cancer question will linger—and confuse—people for years. While there is no proof of adverse health effects, respected lab studies have shown cellphones do affect the brain. A 2011 study that placed active cellphones by people's ears for 50 minutes found participants' brain activity sped up, even though the phones were silent. This activation manifested as increased metabolism of glucose, a sugar the brain consumes, in the areas nearest the phone's antenna. The study, which was conducted by the National Institutes of Health, essentially showed that the brain is sensitive to RF energy, but it did not determine whether this sensitivity is detrimental to the user's health.

Brain cancer's long incubation period is another reason fears persist. Diagnosis and death rates, both globally and in the United

States, have changed little (when adjusted for age) in the three decades cellphones have existed. Brain cancer remains a rare disease. The National Cancer Institute at the National Institutes of Health estimates there will be about 23,000 new diagnoses for all of the United States in 2014. However, cellphones have been popular for only about 20 years, and most health studies have looked only at periods shorter than 10 years. The COSMOS study will fill this gap, but its long-term results won't be available until after 2030.

Cellphone health studies tend to end with the same cautionary note: "More research is needed." While even the scientists spearheading cellphone health research admit that this has become a "tired refrain," they also note it "fully applies in this instance."[11]

SMARTPHONE ADDICTION

Cancer scare or no, smartphone users remain attached to their phones. But how much smartphone use is too much? Smartphones have already altered the way people interact with each other. In her 2011 book *Alone Together*, Massachusetts Institute of Technology professor Sherry Turkle talks about the "world of continual partial attention."[12] It's what happens when mobile technology invades conversations and meetings, stealing people's focus away from their partners, friends, colleagues, parents, and children. Global studies have estimated people check their cellphones every six and a half minutes, on average. Assuming a 16-hour day (with an 8-hour break for sleeping), that means people look at their phones 150 times a day. In 2013, the *New Yorker* satirized this trend in a cartoon depicting a wedding ceremony in which the entire wedding party is preoccupied with their smartphones. Immersed in her phone, the bride mutters to the officiant: "Huh? Oh, yeah—I do."[13]

Simply taking away people's phones isn't a solution. In 2008, the British market research firm YouGov coined a word for the intense anxiety people feel when separated from their cellphone: nomophobia, referring to the fear ("phobia") of being without ("no") our mobile phones ("mo"). And people have been joking for years about the need to use or check their smartphones as a form of addiction.

For many, being without their phones is an experience akin to withdrawal, but the medical community has not formally recognized smartphone or cellphone addiction as its own mental disorder. "It is very difficult to determine at what point mobile phone use becomes an addiction," wrote a British researcher in a 2013 analysis of cellphone usage among Spanish teenagers published in the British journal *Education and Health*. "We are not yet in a position to confirm the existence of a serious and persistent psychopathological addictive disorder." [14]

Mental health professionals who believe smartphone overuse is its own disorder consider it a problem of impulse control similar to compulsive gambling. Dr. Kimberly Young, a psychologist who has been studying Internet addiction since 1994, deems smartphone addiction a form of that condition, because "the issue is not the device. . . . [I]t is what [people] are doing online."

Young founded the United States' first hospital-based treatment center for Internet addiction in 2013 at Bradford Regional Medical Center in northern Pennsylvania, and she defines Internet addiction as "any online-related, compulsive behavior which interferes with normal living and causes severe stress on family, friends, loved ones, and one's work environment. . . . It is a compulsive behavior that completely dominates the addict's life." [15]

To determine whether someone is addicted to the Internet, Young uses a diagnostic questionnaire she created back in the 1990s. Questions include:

- Do you repeatedly make unsuccessful efforts to control, cut back, or stop Internet use?
- Do you jeopardize or risk the loss of a significant relationship, job, educational, or career opportunity because of the Internet?
- Do you use the Internet as a way of escaping from problems or of relieving . . . feelings of helplessness, guilt, anxiety, depression? [16]

The questionnaire has become one of the world's most frequently used Internet addiction tests, and Young also uses it to diagnose smartphone addiction. "Other tests and quizzes have been developed to address new [online] trends," she says, "but I see this as a waste of energy, [because] if a person meets [the Internet addiction test] criteria, they are addicted—it doesn't matter to what."

To devise her questionnaire Young adapted the criteria used to identify pathological gambling as outlined in the fourth edition of the American Psychiatric Association's *Diagnostic and Statistical Manual of Mental Disorders (DSM-IV)*, the authoritative guide health care providers use to diagnose mental disorders. Internet addiction is not classified as a mental illness in the *DSM-IV*, which was published in 1994 and updated in 2000. But the manual's fifth edition, which was published in 2013, does list Internet Gaming Disorder in a special section titled "Emerging Measures and Models," where it describes the disease as "persistent and recurrent online activity [that] results in clinically significant impairment or distress."[17] In a 2012 CNN interview, Charles O'Brien, a University of Pennsylvania psychiatry professor who chaired the Substance-Related Disorders Work Group for the *DSM-5*, explained what kind of work would be necessary to get Internet addiction listed in a future DSM: "We would need studies done in multiple sites. People would have to get together and decide on criteria for the diagnosis. There are . . . a significant number of American therapists who are treating cases like this, but they are generally writing up studies as clinical experience. This is not evidence. You have to do careful studies."[18]

Although smartphone addiction is not a formally recognized disorder, Young says it is "more problematic" than other forms of Internet addiction, because mobile devices make the Internet easily accessible and are difficult for other people—psychologists, parents, spouses—to monitor. Smartphone-based social media, gaming, online gambling, and texting can all be addictive, she says.

Young has privately treated thousands of people for Internet addiction since the 1990s. She typically meets patients for three months of weekly sessions and uses a three-phase process she developed that combines cognitive behavioral therapy with harm-reduction therapy. Since the Internet is a core part of most people's daily lives, Young focuses on promoting "moderated and controlled" usage of the Internet rather than total abstinence.[19] She says she uses the same treatment method for smartphone addiction, and the Internet addiction inpatient program that she developed for the Bradford Regional Medical Center also follows this method. Patients spend the first three days of that ten-day program without access to any digital devices, then "learn to use technology in responsible ways that add to [their lives]."[20]

Young is one of the global experts on Internet addiction, but South Koreans, who have a long history of monitoring and treating it, are now leading the way in smartphone-specific addiction research and treatment. Korea was the first country to open a national Internet addiction treatment center, in 2002, and it now spends about $10 million a year running a nationwide network of treatment centers. The dangers of Internet addiction were plainly apparent in some recent Korean events. In 2009, a Korean couple let their three-month-old baby starve to death because they were immersed in an online 3D fantasy role-playing game. Around the same time, two young Korean men separately killed their mothers after they criticized them for online game playing. People have died after multiday Internet gaming marathons—in China and Taiwan, as well as in Korea. In 2011, Korea instituted a shutdown law that locks people younger than 16 years old out of gaming websites from midnight to 6:00 A.M. to discourage minors from indulging in online gaming binges.

Though smartphone addiction has yet to produce such extreme results—and may never—Koreans are some of the world's most enthusiastic smartphone users, and the country's leaders appear more concerned about smartphone addiction than those of other countries. Korean government agencies regularly conduct studies and surveys on the issue and in 2013, Korean media quoted government officials saying, "Children's addiction to the Internet and smartphones is becoming serious,"[21] and "We felt an urgent need to make a sweeping effort to tackle the growing danger of online addiction . . . especially given the popularity of smart devices."[22]

Korean doctors have defined smartphone addiction as users' unwanted reliance on the devices, and they say it "manifests tolerance, withdrawal symptoms, and dependence, accompanied by social problems."[23] In 2013, the Korean government reported that more than

18 percent of Korean teenagers were addicted to their smartphones, up from 11 percent in 2012. The percentage of afflicted Korean adults also increased year over year, from about 7 percent in 2012 to around 9 percent in 2013. One government survey, which polled students in the fourth, seventh, and tenth grades, found that the number of students who exhibited signs of smartphone addiction was more than twice the number of students who appeared at risk of Internet addiction. Another 2013 Korean government report said smartphone addicts spend seven hours a day browsing the Web, playing games, and sending messages—almost twice the amount of time average users spent on the same activities. Korean research found that online messaging apps are most responsible for keeping "addicts" glued to their phones, followed by mobile games.

The Korean government is particularly worried about smartphones' effects on young people, their educations, and their futures. Korean doctors have reported rising rates of "digital dementia"[24] in teenagers and blamed the overuse of smartphones and other technologies for hindering young people's brain development, including their ability to recall information. In May 2013, one in three Korean middle- and high-school students admitted that excessive smartphone use had caused their grades to fall. A similar percentage said they had tried to cut down their smartphone usage but failed. Almost half of the surveyed students said they could not live without their smartphones.

In 2013, the Korean government said it would require schools to teach students as young as three how to limit gadget use and Internet time. "Young people's smartphone use has increased rapidly in recent years and, truthfully speaking, the government's response has been insufficient," explains Hwang Tae-hee, a representative of Korea's Ministry of Gender Equality and Family. Hwang says the government was partly held back by the lack of smartphone-specific addiction research, so her agency, which helps run Korea's Internet addiction program, is surveying young people about their smartphone use and will use the data to develop a counseling manual.

With support from the Korean Ministry of Health and Welfare, Korean researchers also modified and expanded Young's questionnaire to create the world's first smartphone addiction scale (SAS). The diagnostic test consists of 33 statements, including:

- I feel impatient and fretful when I am not holding my smartphone.
- I feel that my relationships with my smartphone buddies are more intimate than my relationships with my real-life friends.
- I have tried time and again to shorten my smartphone [usage] time, but [keep] failing.[25]

Responses range from "strongly disagree" to "strongly agree" and are graded on a point basis. The researchers hope the creation of an SAS will "serve as an opening for the clinical diagnosis of smartphone addiction."[26]

The Developing World

Manu Prakash wasn't expecting to be shocked when he dropped by a rural clinic during a vacation to central India. Prakash was born in India, and he attended college there before moving to the United States to pursue a master's degree in applied physics at MIT. But Prakash was surprised when a medical student in the village of Sevagram told him about the area's high rate of oral cancer.[27] And he was shaken when the student showed him images of the untreatable lesions inside these patients' mouths. It was too late to save the majority of the patients.

Oral cancer is the most prevalent cancer among Indian men, primarily due to the consumption of cheap chewing tobacco. If oral cancer is diagnosed early, patients have an 80 percent to 90 percent survival rate. If it is discovered in its late stages, the morbidity rate is high. Dentists can identify oral cancer lesions, but Prakash says that rural areas in India typically have only one dentist for every 250,000 people. The dearth of high-quality, standardized medical imaging in rural India further hampers treatment. Lacking funds for imaging machines, health workers may resort to sketching diagrams of their patients' mouths by hand to document their conditions.

Prakash returned to the United States determined to invent an inexpensive gadget that could detect oral cancer early enough for effective treatment. He started building a prototype in 2011 after moving to Stanford University to teach bioengineering. Called OScan, for oral scanner, the system consists of a mobile app, a circuit board with two rows of fluorescent light–emitting diodes (LED), plastic housing that fits over the circuit board, and a plastic mouthpiece with a mechanical track that connects to the housing. A health worker could check a patient for oral cancer by placing the mouthpiece in the person's mouth and snapping a few photos. The LEDs illuminate the patient's oral cavity and expose any lesions while the track enables the worker to move the smartphone/camera steadily and collect a standardized set of images. The OScan app would then upload the images straight from the phone to a cloud-hosted database, enabling doctors in nearby cities to view the images on a separate, password-protected website and screen for oral diseases. (Images can also be saved and uploaded later, if workers lack access to a stable Wi-Fi or data connection.) James Clements, an engineer who worked on OScan as a Stanford graduate student, says a trained worker can take the necessary photos in less than two minutes and the device will work with most camera-enabled Android phones, would cost only about $5 in terms of materials, and can be sterilized after each use so that it can perform hundreds of scans before the mouthpiece needs replacing. Though OScan is still in development at Stanford, Clements says some small, preliminary clinical studies are planned for the United States and in India.

OScan represents one way smartphones can revolutionize health care in the developing world. A few other researchers are also enabling small rural clinics to engage in telemedicine, i.e., health care over long distances via computers, phones, and the Internet. Harvard Medical School researchers have paired iPhones with a $4.99 video camera app and a commonly used ophthalmic lens to capture high-quality images of patients' retinas for the diagnosis of eye diseases. Compared to the cameras that ophthalmologists normally use, which cost tens of thousands of dollars, the iPhone setup is "relatively simple to master, relatively inexpensive, and can take advantage of the expanding mobile-telephone networks for telemedicine," the researchers wrote in the *Journal of Ophthalmology*.[28] Australian medical students have created a low-cost, easy-to-use, smartphone-based system that can detect pneumonia. The system, which consists of an Android and Windows Phone app called StethoCloud and a digital StethoMic stethoscope that plugs into the phone's audio jack, records a patient's breathing, performs basic offline analysis on the patient's respiratory rate, and then uploads the data to the cloud for deeper analysis.

Smartphones are also helping health workers access medical guidelines for rapidly and accurately assessing patients. Health workers in Malawi's southern districts Blantyre and Chikwawa are using Android phones to evaluate children who have meningitis. With support from the Britain-based Meningitis Research Foundation, five health centers in Blantyre and two in Chikwawa have deployed an app that lists the World Health Organization's Emergency Triage Assessment and Treatment protocol for meningitis. Health workers consult the app when sick children are brought in to the centers to help them decide which children need to see a doctor immediately, which get priority to see a doctor soon, and which can wait. The app also generates an ID number that can be used to track children who are sent to hospitals, to ensure that they receive proper care. D-tree International, the Massachusetts-based health nonprofit that created the app, says more than 120,000 cases have been logged through the app since it was released in December 2012.

In these circumstances, smartphones can be lifesavers. "Smartphones have great potential for more advanced, two-way types of communications and for health professionals who are engaging with different populations in real time," says Julie Pohlig, chief strategist at the California-based Vital Wave Consulting, which focuses on emerging markets. "In health care it can be highly advantageous to have the advanced cameras and video and diagnostic tools that are available on smartphones."

International aid and development organizations started using PDAs and cellphones in their outreach projects as early as the 1990s. Joel Selanikio was one of the first epidemiologists to experiment with mobile devices. "Back in 1995, I was a young outbreak investigator for the CDC [the U.S. Centers for Disease Control and Prevention], and exploring the use of things like the Apple Newton, and later the Palm

PDAs, for data collection," he recalls. "Not much was being done at the time, but we used [Palm OS] software . . . to put forms on PDAs."

In 2001, the charity Save the Children deployed handheld computers loaded with Microsoft's Pocket PC software to survey remote Nicaraguan communities about nutrition and other topics. Collecting health-related data on PDAs was faster and more accurate than paper records, but obtaining and maintaining the devices in developing countries was challenging. As Selanikio wrote in a 2013 blog post: "Unfortunately, PDAs were aimed at the rich world market of businesspeople and other professionals [and] in order to use them for a health survey in Namibia, for example, the difficult and expensive burden of getting them to Namibia was placed squarely on the shoulders of the global health system. . . . This was too heavy a burden to bear: the PDA-based data collection activities typically collapsed after just a few years." [29]

The other drawback to using handheld computers for international development work at that time was the inability to transmit data from the field. As a 2001 Reuters article explained, when Save the Children wanted to upload data from its Nicaraguan aid workers to its network, the charity had to "whisk" the devices "by motorcycle through the jungle" to its offices. [30]

Though the PDA projects flopped, cellphones later became a mainstay of development work, because governments and development organizations didn't need to import them and data could be uploaded and downloaded in the field, since networks had improved. Today there are mobile health programs that provide health care over phones, mobile money programs that give the poor access to financial services via their phones, mobile agriculture programs that connect farmers to useful information through their phones, and mobile education programs that teach people skills, such as English, over their phones.

Many of these projects assist people who live on less than $2 a day. Some outreach projects donate phones to poor communities, while others rely on local people to use their own. Since few people in those communities own smartphones, the latter type of project usually focuses on basic cellphones. The GSMA estimates that only 8 percent of people in the developing world own smartphones, and that figure dips lower in rural areas, where network coverage issues, battery life, and literacy barriers—both reading and technological literacy—further limit their adoption. Smartphone ownership in rural Africa, for example, is estimated to be about 1 percent.

Yet as technology advances, development organizations are keeping pace by investing in smartphones and apps. Experts say smartphone outreach projects can offer greater gains than basic phone projects if they are carefully planned and executed. "Smartphones have a higher cost but also offer higher opportunity," says Patricia Mechael, the executive director of the mHealth Alliance, a Washington, D.C.–based nonprofit that aims to advance mobile health technology globally. "Health outcomes can be greater with smartphones, as can [data] efficiencies."

Picking up where the PDA projects failed, one of the most effective ways smartphones are being deployed for public good is the collection and transmittal of public health data via Android-based apps. Smartphones are a natural choice for these tasks, because they enable workers to record, analyze, and share data more easily than paper records. Laptops and tablets offer similar capabilities, but they are bulkier and more expensive.

SEND-Ghana, an Accra-based NGO, says smartphone-based surveys are helping it reduce health care inequality. To achieve this goal, SEND closely monitors the National Health Insurance Scheme (NHIS), a program the Ghanaian government established in 2003 to provide universal health coverage to its citizens.[31] In 2011, SEND recruited hundreds of volunteers to interview residents of Ghana's 50 poorest districts about their ability to access NHIS services. The group provided its volunteers with Samsung Android phones to record interview responses in a custom-made survey app. The smartphones yielded more accurate data and enabled faster reporting than previous methods. Analysis was quicker, too, because the smartphone data could be directly exported to spreadsheet software for study. After joining forces with other NGOs, SEND was able to use that information to successfully lobby the Ghanaian government to improve access to NHIS services.[32]

Smartphone-based development projects have gotten a boost from software tools that let organizations create their own, locally relevant apps without having to construct and maintain the infrastructure and back-end support systems necessary to collect, store, and analyze masses of data. A Ghanaian developer built the SEND app using free, open-source Android software called Open Data Kit Collect, which is partly financed by Google and maintained by researchers at the University of Washington in Seattle. Other global development organizations have written apps based on software from Akvo, a Dutch nonprofit, and DataDyne, a Washington, D.C.–based social enterprise run by Selanikio, the trailblazing epidemiologist, technologist, and social entrepreneur. Workers and volunteers in Africa and other regions are using these apps to conduct field surveys, audit health facilities, and gather information about water and sanitation conditions.

7

Privacy

Privacy is one of the most pressing issues for smartphone users today. Companies are leveraging apps to snoop on people's travels, habits, and purchases. The government has spent years collecting calling records and other smartphone data about large numbers of people. Law enforcement agencies are increasingly accessing smartphone information not just on criminals but on law-abiding citizens. To curb these intrusions, users need to understand the many different ways they are being monitored.

Apps are a major reason why smartphones are privacy invaders. In 2012, Carnegie Mellon University (CMU) researchers analyzed the 100 most popular Android apps and discovered that more than half gathered user data, including contact lists, user locations, and unique device IDs assigned to each phone.[1] Device IDs are long letter-and-number codes that Apple and Google give to each phone so it can be distinguished from other iPhones and Android phones. On iOS, these IDs are called "Unique Device Identifiers" (UDIDs) and, on Android, "Android IDs."

Android apps can also read a user's phone number, Gmail address, full name, calendar data, call log, Web-browsing history, and bookmarks. iOS apps can look at users' calendar, reminders, and photo library, and access their Facebook and Twitter accounts. Some apps must collect some of these user data to operate effectively. For example, apps that help people find local deals, navigate public transportation, and detect where they parked their cars all need to know their users' locations. iPhone developers use UDIDs to give beta users early access to their apps, before they go live in the App Store, and Apple uses them to link people's iTunes login credentials with their iOS devices, which is useful when you need to redownload an app you've already paid for.

But many apps collect data that have little relevance to their functionality or purpose. For example, the CMU study discovered that Brightest Flashlight Free, an app that turns a smartphone's camera flash into a flashlight, gathered location data from its users. So did the game Angry Birds. Pandora, the Internet radio service, collected users' contact lists. CMU professor of computer science Jason Hong says some of the findings jolted people. "When you start hearing that that game would need your location data or this flashlight app needs your unique ID, . . . it doesn't seem reasonable," he explains. "Whenever I give talks, people in the audience start uninstalling apps [from their phones] right away."

Developers routinely gather this nonessential user information so they can send it to advertisers. The CMU researchers found that 61 of the 100 apps in their study did this. Angry Birds gave sensitive user data to seven mobile ad companies. So did Brightest Flashlight.

Users typically don't know what an app does with the data it collects. CMU researchers uncovered these connections after they read the apps' code and found small chunks of code from third-party advertising "libraries" embedded in it, says Hong. Mobile ad networks create these libraries and provide them to developers to stick in their apps. Developers incorporate the code so they can show ads in their apps, but integrating the libraries also enables ad networks to access user data.

Mobile ad companies often leverage this information to group app users into audience categories or segments that can be targeted. Flurry, a San Francisco–based app analytics and advertising firm, is one of these companies. Flurry partners with developers who enable it to gather data about smartphone users across hundreds of thousands of apps, and it sells advertisers access to those consumers by grouping them into more than 40 different "personas," including "Singles," "High-Net Worth Individuals," and "LGBT."[2] Advertisers will typically pay a premium to reach their desired audiences, which means more money for companies like Flurry and the developers who partner with them. For developers, the more they know about their users, the more valuable their app is to advertisers.

Device IDs facilitate this targeting by enabling companies to track smartphone users across apps and devices. The companies assume each device ID represents one person and then gather and save user information based on the ID. This practice has rankled civil liberties advocates and consumers and triggered several lawsuits in which plaintiffs alleged Apple let apps collect their personal information without their consent. Judges eventually dismissed the suits, but Apple recognized the need for a different type of device ID specifically designed for advertising use. Advertisers can still trace consumers in iOS apps via the new IDs, called "Identifier for Advertisers" (IDFAs or IFAs), but iPhone users can opt out by selecting a limit ad tracking option in their phone settings. iPhone users can also reset their advertising IDs by pressing a command in their phone menus. Google has also updated Android to include new IDs (called, simply, "Advertising IDs") that function in the same way.

When Google announced the new IDs, the company said the technology would "give users better controls" while providing developers with a "simple, standard system to continue to monetize their apps."[3] Hong is skeptical. "On the surface, advertising-specific IDs appear to be a really good idea," he says. "But I'm concerned that people will try to find a workaround." Hong draws an analogy to the desktop Internet where Do Not Track initiatives spurred advertisers to devise tracking technologies that are harder for users to detect and delete. He predicts that "we will see the same challenges with smartphones as with [the desktop Internet]."

Though apps gather far more data than most people realize, they don't do it without warning. Android apps are permission-based, meaning they tell users what data they plan to collect and request their permission. This information appears as soon as a smartphone user tries to install an app. iOS apps don't request permissions during installation, but users are notified the first time an app tries to access sensitive data, and they can also disable or enable specific permissions for individual apps in their phone settings.

Hong thinks Apple and Google "tried their best to figure out" how to notify people of app permissions, but he says both the iOS and Android briefing systems have weaknesses. The iOS approach, he says, is "good, in that an app tells you when it's about to do something [that requires permission], but it's not always clear why it's going to be doing it." With Android's approach, "you have to hit the install button before you see the permissions," which may not be an effective way to present the information, because "at that point, you've already chosen to install the app." Android also takes an all-or-nothing attitude toward permissions, so a person who wants to use a certain app must either agree to all the requested permissions or decide not to download and use the app at all. Hong says both Android and iOS wrestle with "how to explain all the things going on inside an app without having too little or too much detail."

Recent research shows that Android users don't read and/or don't understand app permissions. A 2012 University of California–Berkeley study of Android app permissions found that only 17 percent of study participants (all of whom were Android users) looked at the app permissions when they installed apps from Google Play. Even when they did, comprehension of the permissions was low. The researchers gave participants a multiple-choice questionnaire that asked them to describe the meaning of various permissions. Based on the responses, less than 25 percent of the participants exhibited "competent" comprehension of app permissions.[4]

In a written summary of their findings, the researchers concluded, "The current Android permission system does not help most users make good security decisions."[5] They cited "warning fatigue" and confusing descriptions as two of the system's major flaws.[6] In other words, smartphone users see so many permissions warnings that they stop paying attention. App permissions are also written in technical jargon and fail to specify the purpose of their requests. For example, many apps ask permission to read a user's "phone status and identity." People don't realize it, but granting that permission enables the app to detect whether the phone is actively making or receiving a call and lets it copy the phone's IMEI [International Mobile Equipment Identity] number—another type of unique identifier phone makers assign to the phones they manufacture, so they can be identified in case of theft. Apps don't need users' IMEIs to function, so developers that solicit them are likely sending that data to ad networks for ad targeting.

Apps are proliferating so rapidly with so little regulation that former EFF technologist Dan Auerbach refers to the app market as "the Wild West." "The barrier to entry to becoming a developer is so low, most developers don't have much experience handling user data," he explains. "Users are at the whim of developers, which is scary."

Developers, in turn, are at the whim of advertisers, and that is the root of many app privacy problems. The freemium app economy partly relies on the sale of consumer data and ads to make money. In 2013, Hong and his researchers interviewed developers in the United States and surveyed developers worldwide about their attitudes toward privacy. He found that developers who invade user privacy often do so unintentionally: "Sometimes developers don't understand what these advertising libraries are doing. They know it's an advertising library, and they throw it in their apps, but they don't know how often data is collected, where it's being sent, or the range of data. They just think, 'My app won't work [or make money] if I don't put in this library.' . . . I'm pretty sure advertisers don't see themselves as bad guys [either]. They're trying to figure out how to offer free apps and have innovation. The challenge is, once your business model is predicated on ads, you have a strong incentive to collect as much data as possible about [users] because of behavioral ads. If developers can double their [ad] click-through rates based on having more user information, and that would essentially double their revenues—well, you can see how that whole cycle goes."

Disrupting that cycle will require better developer education, guidelines, and tools, says Hong. "In our interviews we did see developers who said they wanted to do something about privacy, but it was not clear to them what they should be doing."

In the absence of a specific app privacy law, the U.S. Federal Trade Commission (FTC) has gone after a few app developers for threatening consumer privacy and violating the FTC Act, which prohibits unfair or deceptive business practices. In February 2013, Path Inc., the start-up behind the eponymous social networking app, settled FTC charges that it had violated the FTC Act by collecting users' smartphone address books "without their knowledge and consent."[7] The FTC said Path also allowed approximately 3,000 preteen children to sign up for its service without parental consent, which is a violation of the Children's Online Privacy Protection Rule (COPPA). Updated in 2013, the COPPA rule requires apps (and websites and other online services) to obtain "verifiable parental consent" before collecting personal information—such as names, addresses, contact information, photos, and videos—from children younger than 13.[8]

The FTC also went after the creator of the Brightest Flashlight app that had rattled people in the CMU privacy study. In December 2013, Idaho-based Goldenshores Technologies settled charges that it had violated the FTC Act by "deceiv[ing] consumers about how their geolocation information [and unique device IDs] would be shared with advertising networks and other third parties."[9] Among other stipulations, the FTC settlements required both companies to delete the personal information they had gathered and improve their disclosure and permissions procedures. Path also had to pay an $800,000 penalty for its COPPA violations.

In February 2013, on the same day it announced the Path settlement, the FTC issued a mobile privacy report that recommended that platform providers educate developers on privacy issues, require developers to make privacy disclosures, and "reasonably enforce" various best practices.[10] The guidelines are nonbinding, but the agency said it would "view adherence to [strong privacy codes] favorably in connection with its law enforcement work."[11] The GSMA, the Commerce Department's NTIA agency, and the California attorney general have also formulated "best practice guidelines" for developers. Hong has discussed app privacy with the FTC, and he doesn't think a specific app privacy law is necessary, especially since it's not clear what the law should be, but he contends that greater transparency is needed to keep the app business healthy. He says: "We've got this interesting ecosystem, with a gigantic number of apps, and we don't want to destroy it, because we're seeing lots of innovation. People are generally OK with ads, and behavioral ads, if things aren't hidden and they can make an informed decision. We just need a better conversation in the public sphere as to what's going on." Given the power that smartphone platform providers wield over their app stores, strengthening app privacy may depend on companies like Apple and Google instituting stricter rules.

RETAILER AND CARRIER TRACKING

Apps aren't the only way companies peep on smartphone users. Retailers use smartphone Wi-Fi and Bluetooth connections to surreptitiously learn about consumers who visit their stores. They do this by purchasing technology that grabs IDs known as Media Access Control (MAC) addresses, which are 12-digit alphanumeric codes that help smartphones communicate with Wi-Fi networks. All Wi-Fi–capable smartphones have unique MAC addresses assigned by their manufacturers. They also have separate Bluetooth MAC addresses, which are used to pair smartphones with Bluetooth accessories, such as wireless headsets.

Retailers can read people's MAC addresses through their store Wi-Fi networks or special sensors. The technology providers are supposed to de-identify this information and retailers primarily use it for general insights, such as which store areas attract the most and least traffic and how long shoppers browse and wait in line before making purchases; but over time, retailers could gather enough data about shoppers to create more detailed, though anonymous profiles. To protect their privacy, people can turn off their smartphones' Wi-Fi and Bluetooth connections before entering stores or alter their MAC addresses but the latter requires hacking their phones to obtain access to all of its files and programs—a process that is called "rooting" when done on Android phones and "jailbreaking" when done on iPhones.

Senator Charles Schumer (D-NY) and the Future of Privacy Forum (FPF), a Washington, D.C.–based think tank, have publicized the need for the responsible use of these so-called smart store technologies. In October 2013, the FPF released a code of conduct for the firms that provide this technology. The companies agreed to ask their retailer partners to put signs in stores telling shoppers they are being tracked. The signs would also give shoppers the address of SmartStorePrivacy. org, an FPF website where they can opt out of being monitored. Civil liberties advocates, such as the EFF, applauded this attempt toward self-regulation but also highlighted the policy's limitations, including the fact that retailers can decline (and so far have declined) to post the proposed signs.

Carriers are tracing smartphone users' shopping patterns, too—

and they're selling the information. Verizon, Sprint, and AT&T recently established programs that provide data about their subscribers to marketers, advertisers, and retailers. The three carriers mention these programs in their subscriber privacy policies, and they say the data is made anonymous. Verizon, Sprint, and AT&T include their cellphone subscribers in these programs by default. Subscribers must take the initiative to opt out.

Verizon's program is called Precision Market Insights. Launched in October 2012, it collects "detailed demographics and geographics," including "shopping habits, interests, travel patterns, and mobile browsing trends,"[12] for all of Verizon's consumer subscribers—i.e., not its corporate or government subscribers—unless they opt out. Verizon gathers this data from its smartphone users by recording their locations, the websites they visit on their phones, including the terms they use for mobile searches, and how they use their apps. In the next step, as a 2013 CNN story explained, "Verizon sends that data to an internal database, matching it up with a deep trove of demographic information about you from companies including data giant Experian. The data is stripped of any personally identifying information, aggregated into categories, and are placed into reports for Precision customers to use."[13] For example, a sports team could request data on the demographics (i.e., income level, age range) of the people who attend its games and use this data to tweak its local advertising.

Sprint introduced its Reporting & Analytics program at the same time as Verizon. Like Verizon, Sprint gathers two general types of subscriber data: mobile usage information and consumer information. Mobile usage information is data Sprint collects through its network, such as location information and the websites that subscribers visit and the types of apps subscribers use on their phones. Consumer information is more specific to Sprint products, services, and subscribers, such as the data and calling features its subscribers use, how much or often they use them, and their gender and age range. Sprint uses the data to produce business and marketing reports or sells the data to other companies to produce reports. As examples, Sprint says;

We may aggregate customer information across a particular region and create a report showing that 10,000 subscribers from a given city visited a sports stadium. . . . [W]e may share de-identified location data . . . to create a report showing that 10,000 mobile subscribers passed a retail location on a given day.[14]

AT&T has a similar External Marketing & Analytics program that sells customer location information, mobile browsing habits, and app usage to retailers, TV networks, and device makers. The example AT&T employs is similar to Sprint's. The carrier says it could create a report that shows a retailer the number of AT&T wireless devices in or near its store locations by time of day and day of the week, along with general shopper demographics, such as age range and gender. Though that type of report doesn't sound intrusive, civil liberties advocates regard AT&T's program as a greater potential privacy threat than Sprint's or Verizon's, because AT&T is able to aggregate detailed usage data across its wireless, Wi-Fi, and Internet Protocol TV (IPTV) networks.

All of these programs are designed to manage privacy issues by synthesizing information without ever identifying specific individuals. But as a 2013 *MIT Technology Review* article noted:

The concerns about making such data available . . . are not that individual data points will leak out or contain compromising information but that they might be cross-referenced with other data sources to reveal unintended details about individuals or specific groups.[15]

Recent events and research have shown that data tracking can have worrying implications even if people aren't identified personally.

THE NATIONAL SECURITY AGENCY CONTROVERSY

Murky consumer-tracking and data-mining initiatives validate the need for smartphone privacy protections, but for many smartphone users it was the NSA's phone-records scandal that clarified those points and made smartphone surveillance feel like a real, immediate concern.

The NSA controversy began in June 2013, when whistleblower and former NSA contractor Edward Snowden leaked a stash of classified NSA documents to media organizations. The resulting articles, published in the British daily the *Guardian* and elsewhere, revealed that the NSA was keeping records on millions of American landline and cellphone calls every day.[16]

President George W. Bush authorized the program in 2001 following the September 11 terrorist attacks. It operated without court supervision until 2006, at which point the NSA began using secret court orders to compel AT&T, Sprint, and Verizon to tell it what numbers their subscribers were calling or receiving calls from, at what time, and for how long "on an ongoing daily basis."[17] The call data did not list people's names or addresses, but it did contain information that could be used to determine a person's identity, such as the caller and call recipient's phone numbers and IMEI and IMSI (International Mobile Subscriber Identity) numbers, the latter of which is stored in a phone's SIM card and used by carriers to identify subscribers on their networks. The NSA saved this carrier-supplied information in databases for up to five years and searched it several hundred times a year.

Officially, the NSA was only supposed to query its phone records when it had "reasonable, articulable suspicion" that the phone numbers it was researching were associated with "specific foreign terrorist organizations,"[18] but media reports found instances when NSA employees strayed beyond this rule, mostly unintentionally. NSA phone data queries could also be very wide in scope, with any one potentially involving thousands of Americans' phone records.

People believed the NSA tracked all American phone activity until February 2014 when the *Wall Street Journal*[19] and *Washington Post*[20] reported that technical and compliance barriers had limited the agency's data collection to 30 percent or less of all U.S. calls (landline and wireless combined). Verizon, for example, hands over data regarding only its landline calls, not its cellphone subscribers, and T-Mobile does not automatically share any data with the NSA. But privacy advocates have noted that the agency still collects information in bulk for the phone-records program even if it is not comprehensive.

The revelations about the NSA's bulk-phone-records and other data-collection programs, including ones involving people's online communications, such as e-mail, triggered more than a dozen legislative proposals, several lawsuits from civil liberties groups, and a number of mass rallies. Each of the initiatives sought to curtail governmental monitoring of private communications.

To quell the uproar, President Barack Obama requested the creation of a review group to assess U.S. intelligence policy, including the NSA's data-collection practices. The five-person advisory panel made more than 40 recommendations in a December 2013 report, including the substantive modification of the NSA's bulk-phone-records program, because the government lacked "sufficient justification" to collect and store the sensitive data.[21] The report also said the bulk-phone-records program had not proved to be "essential to preventing [terrorist] attacks."[22] A few weeks later an independent federal watchdog agency, the Privacy and Civil Liberties Oversight Board, met with the president and told him the majority of its members had concluded that the bulk-phone-records program was ineffective and "lack[ed] a viable legal foundation,"[23] and therefore should be shut down. Obama declined to do that, but in January 2014 he announced program reforms, including stricter legal requirements for NSA bulk-phone-records queries, limits on the scope of accessible data, and the development of a new phone-records program to replace the government's controversial practice of storing massive amounts of American phone data on its own servers. In his speech, Obama called the bulk-phone-records program a "powerful tool" that can help intelligence agencies "identify patterns or pursue leads that may thwart impending threats."[24] He also acknowledged that "the government collection and storage of such bulk data . . . creates a potential for abuse."[25]

In March 2014, Obama recommended that the government leave these data with the carriers and submit search requests to them when needed (subject to judicial approval.) Civil liberties advocates applauded the idea of ending the government's systematic collection of Americans' phone records, but also noted that the proposal would leave "the underlying legal theory that spawned [the program] intact."[26]

The NSA scandal showed how vulnerable phone-related information is to government intrusion and the complicated role that carriers and other service providers play as keepers of that information. The Fourth Amendment safeguards people's "papers and effects"[27] against unreasonable search and seizure by requiring a warrant, but the government doesn't consider phone data subject to the Fourth Amendment. Under the third-party doctrine, a legal concept established decades ago, information knowingly revealed to a third party loses its Fourth Amendment protections. From the government's perspective, individuals who sign up for cellphone service automatically surrender their right to privacy, because they know their carriers will track their phone activity and keep records of it. In fact, the government often refers to this data as the carriers' "business records."[28]

Judges have upheld the third-party doctrine, although one federal judge, U.S. District Court Judge Richard Leon, disagreed with the government's interpretation when he ruled on a lawsuit that challenged the NSA's bulk-phone-records program. In his December 2013 written opinion, Leon said that the program was "almost-Orwellian" and likely constituted "an unreasonable search under the Fourth Amendment,"[29] which would make it unconstitutional. His decision granted the case's two plaintiffs—a conservative legal activist and one of his clients—a preliminary injunction that would have blocked their phone data from government monitoring. But Leon put his order on hold in anticipation of a government appeal, which the Justice Department soon requested.

The NSA controversy also illustrated how the law distinguishes between content and metadata and how that distinction jeopardizes smartphone privacy. Metadata is the transactional data surrounding a technological action, such as the date and time a phone call was placed. Content is the substance of the action, such as the audio portion of a call. The data the NSA compiled in its bulk-phone-records database was metadata, not content. The government typically does not view metadata as sensitive information, which is problematic for smartphone users, because they create metadata each time they make and receive calls, send and receive e-mails, use apps, search and surf the mobile Web, and take photos. As Hanni Fakhoury, an EFF staff attorney who specializes in privacy issues, says, "Smartphones can generate big data trails."

Civil liberties advocates say laws shouldn't treat metadata and content differently. While metadata isn't content in the traditional sense, it reveals who talked to whom at a given point in time. That information can convey details, such as people's political leanings, religious affiliations, and medical conditions that may be more sensitive than the content of a one-minute phone conversation. Law enforcement can use metadata to reconstruct people's past activities and to predict their future actions. Senator Ron Wyden (D-OR), who was an outspoken critic of the NSA's bulk-phone-records program, has called metadata "a treasure trove of human relationship data."[30] Groups, including the EFF, have argued that government collection and analysis of phone metadata threatens Americans' First Amendment–protected freedom of association, because cellphone users may avoid calling certain people (such as an undocumented workers' outreach organization) or places (such as mosques) out of fear that their behavior will be tracked and misinterpreted.

Metadata is also not as anonymous as it appears to be, since some can be matched to people's names using readily available online tools. In the wake of the NSA bulk-phone-records program revelations, Stanford University computer science researchers created an Android app, MetaPhone, that pulled phone call metadata from phones. They asked volunteers to download the app and grant them access to their data. Once they had gathered a data set the researchers randomly selected 5,000 phone numbers and found they were able to match 27 percent of them to people or businesses just by searching for the numbers in Facebook, Google's business listings service Google Places, and the online review site Yelp. The researchers also took a separate group of 100 numbers and identified 91 of them by manually running them through Facebook, Google, Yelp, and public records databases.

In a follow-up study that examined three months of phone metadata from 546 volunteers, the researchers found that 30 percent had contacted pharmacies during that period, 8 percent religious institutions, and 7 percent "firearm sales and repair" businesses. Calls were also placed to Alcoholics Anonymous, a reproductive pro-choice organization, labor unions, divorce lawyers, and sexually transmitted disease clinics. In a blog post, the researchers wrote, "phone metadata is unambiguously sensitive, even in a small population and over a short time window."[31]

PHONE LOCATION DATA AND WARRANTLESS SEARCHES

Both the metadata distinction and the third-party doctrine are factors in the ongoing privacy debate about smartphone location data. Civil liberties advocates consider location records protected under the Fourth Amendment, but since location data is not communications content, and subscribers allow carriers to collect it, many courts consider the information to be carrier property. These courts say police and the government need only a court order—not a search warrant—to request location data from carriers. Court orders are easier to obtain than search warrants, because they do not require police to show probable cause that a crime has been or is being committed. The police only have to claim the information is relevant and material to an ongoing investigation.

Statistics indicate that federal, state, and local law enforcement agencies access the location data of tens of thousands of cellphones and smartphones a year through carriers. After Senator Ed Markey (D-MA) pushed carriers to disclose the number of law enforcement requests for phone-related data they receive annually, eight of the then-largest U.S. carriers revealed that they had collectively responded to 1.1 million requests in 2012. Markey says the figure should be higher, but Sprint "did not provide complete information" in its response.[32] An unspecified portion of those requests sought location information.

AT&T and Verizon supplied more detailed and up-to-date information in early 2014 when they separately published "transparency reports" in response to pressure from privacy activists and others seeking greater clarity on how and when they divulge customer data to law enforcement. Verizon said it received approximately 38,000 law enforcement demands for location data in 2013, through either court orders or search warrants, and that the number of warrants and orders for location information is increasing each year.[33] AT&T said it received nearly as many demands for location data in 2013.[34]

The NSA also collects location data from cellphones worldwide, according to a 2013 *Washington Post* investigation[35] based on confidential NSA documents supplied by Edward Snowden. The NSA's location-tracking efforts are focused on foreign intelligence targets outside the United States. But since the NSA obtains phone location information by tapping into telecommunications providers' "key network routing points,"[36] the agency ends up "incidentally"[37] collecting the locations of an undetermined number of American cellphones, including those of Americans traveling abroad. The NSA saves this location data in a database and uses it to determine social relationships, such as whom its targets might be meeting or accompanying.

The NSA says it does not currently collect Americans' domestic phone-location data in bulk, but it has in the past, and it may resume doing so in the future. After months of obfuscation on the subject intelligence officials revealed this information during an October 2013 Senate Judiciary Committee hearing. After the *New York Times* reported that the NSA had gathered cell tower location information from U.S. carriers, the agency's then-director, Keith B. Alexander and the director of National Intelligence James R. Clapper confirmed it did so in 2010 and 2011 as part of a secret pilot project designed "to test the ability of its systems to handle"[38] such information. (This project was separate from the NSA's phone metadata program.) At the same hearing, the officials said the NSA had never used the location data for intelligence purposes, but Alexander admitted that could change, and that collecting the locations of cellphone calls for NSA analysis "may be something that is a future requirement for the country."[39] The NSA would need to apply for permission to resume location tracking on a bulk level, but the special court that assesses the agency's demands has approved almost all of its requests thus far. The NSA has also said it will inform Congress if it starts gathering phone-location data.

Civil liberties advocates object, since even anonymous location data can be traced back to specific users because phone usage patterns are largely predictable. A 2013 study by researchers from Harvard, MIT, and Belgium's Louvain University found it was possible to "uniquely characterize" nearly everyone in a data set of 1.5 million people using a few cellphone-location data points.[40] After examining 15 months of anonymous user data from a European carrier the researchers were able to differentiate 95 percent of the people with just four randomly chosen data points that disclosed the user's location and time of day. This information could then be cross-referenced with publicly available information, such as a person's home or work address, to identify and track a particular individual, much the way the MetaPhone researchers did with phone numbers in their experiment. The American Civil Liberties Union (ACLU) likes to say, "If the government knows where you are, the government knows who you are."[41]

Senator Wyden has been trying to pass a federal law prohibiting warrantless phone tracking since 2011. His Geolocational Privacy and Surveillance Act (GPS Act), co-written with and sponsored by Representative Jason Chaffetz (R-UT), is stuck in committee review after being reintroduced in Congress in March 2013. If it passes, it would require government entities, including the police, to prove probable cause and obtain warrants before tracing a suspect's cellphone location in real time or acquiring historical cellphone location data. The bill makes exceptions for emergency situations, such as national security threats or tracking people who are in immediate danger.

In the absence of a federal law, states have been taking action. In June 2013, Montana was the first state to enact an anti-cellphone tracking law. Maine passed a similar law a few weeks later and Indiana, Maryland, Utah, and Virginia passed broader laws in March 2014. The Maine and Montana laws mirror the GPS Act though are less extensive while the Utah law requires a probable cause warrant not just for location information, but also for "stored" and/or "transmitted data" from "electronic devices"—language that the ACLU notes could be interpreted to protect a range of electronic communications content.[42] The Indiana, Maryland, and Virginia laws only pertain to real-time (not historical) location information. The Massachusetts and New Jersey Supreme Courts have also ruled that police need warrants to track phones, though the Massachusetts decision specifically applies to requests concerning longer time periods, such as two weeks or more of location information.

A related area of concern is whether the police can search people's cellphones after they have been arrested. Civil liberties advocates say these examinations are overly invasive when they include smartphones. As ACLU technologist and policy analyst Chris Soghoian has written: "The type of data stored on a smartphone can paint a near-complete picture of even the most private details of someone's personal life."[43] To illustrate his point Soghoian published a document that a Department of Homeland Security agency had submitted to court in connection with a 2012 Michigan drug investigation. The document outlined what law enforcement was able to learn after commissioning a forensic exam of a suspect's iPhone. The trove of sensitive information included call activity, contact lists, stored voicemails, text messages, photos and videos, apps, eight different passwords, and 659 geolocation points, including 227 cell towers, and 403 Wi-Fi networks to which the phone had previously connected.

Courts are divided on the constitutionality of warrantless cellphone searches. A legal doctrine known as "search incident to arrest" permits police to search a suspect during or immediately after a lawful arrest. But when the Supreme Court developed that rule in the 1960s it was intended to permit the search of a suspect's clothing and the immediate vicinity for weapons and evidence—not a smartphone that carries far more data than a person's pockets. The Stored Communications Act, the federal law that regulates the government's access to digital information stored by a service provider, shields an individual's text messages, e-mails, and Facebook and Twitter messages from government agencies for 180 days, unless they produce a warrant. But the act was passed in 1986, and it's not clear whether it trumps the search incident to arrest rule in the context of digital information saved on smartphones.

EMERGING PRIVACY THREATS

Technology evolves so quickly that civil liberties advocates worry that smartphone surveillance methods that aren't common today will soon become widespread. The ACLU has warned, "More and more when it comes to monitoring the public, [technological] capability is driving policy. The limits of law enforcement surveillance are being determined by what is technologically possible, not what is wise or even lawful."[44] These technologies include ways to remotely collect smartphone data.

Stingray surveillance systems, which capture phone IDs within a particular vicinity, are an emerging smartphone privacy threat. Stingrays are portable, antennae-equipped devices that mimic cell towers and trick nearby phones into connecting to them so that they can grab the phones' ID numbers. Every cellphone has multiple ID numbers, but the ones Stingrays snare are mostly IMEI and IMSI numbers, which is why some people call them IMSI catchers.

Law enforcement agencies use Stingrays to track a suspect's phone location—and thus, the suspect—in real time. Since Stingrays are as small as a shoe box, they can be placed in a car and transported around a suspect's neighborhood. Once the Stingray picks up the suspect's phone signal, police measure its strength from varied angles and map the data on a computer, until they locate the suspect. The systems are controversial for two reasons: they arguably invade the suspect's privacy and they scoop up the phone data of any person in typically a one-mile radius. Police or the government could then use this data to trace people's actions via their phones and uncover their names. This method could be used to remotely identify attendees at a protest, for example. Police would simply have to activate a Stingray near the protest and collect people's phone IDs.

Privacy groups say these broad searches violate the Fourth Amendment, especially since some Stingrays could enable police to eavesdrop on phone calls through interception modules installed by their manufacturers. These would function like a wiretap but with wider reach and less judicial oversight, since law enforcement generally need only a court order to deploy a Stingray, whereas a wiretap always requires a warrant.

The Department of Justice has admitted that Stingray use is "very common"[45] among federal agents, and Stingray investigations are surprisingly prevalent among local police, as well. The FBI is known to lend its stingrays to state and local law enforcement agencies, and some state and city police have purchased their own, though they are pricy—as much as $400,000 per system, according to one estimate. The Los Angeles Police Department bought a Stingray in 2006 and routinely uses it for burglary, drug, and murder investigations.[46] The Florida Department of Law Enforcement has spent more than $3 million on Stingrays since 2008 and it loans them to regional and local law enforcement agencies, according to the ACLU.[47] Court cases and investigations by journalists and civil liberties advocates have also uncovered local police use of Stingrays in Arizona, Indiana, northern California, and Fort Worth. In December 2013, *USA Today* reported that at least 25 U.S. state and local police departments owned Stingrays, and that some had defrayed the cost through federal antiterrorism grants, even though they used the devices for "far broader" purposes.[48]

Civil liberties groups often discuss Stingrays alongside another cellphone information-gathering tactic called "tower dumps." These are time-targeted information reports related to specific cell towers. Police investigating a crime in a particular location can ask carriers to identify all the cellphones that were connected to the nearest cell towers (for calls, for data, or simply because they were in proximity) during a given period of time—usually a few hours. Like Stingrays, tower dumps can hoover up hundreds or even thousands of law-abiding citizens' location information in pursuit of clues about a suspect or crime. But unlike Stingrays, tower dumps require carrier cooperation, and they can't track people in real time. In their transparency reports AT&T and Verizon said they received about 1,000 and 3,200 tower dump requests, respectively, in 2013. One in four U.S. law-enforcement agencies have used tower dumps to obtain information, according to *USA Today*.[49]

Government and police surveillance aren't the only Stingray-related concerns. Civil liberties advocates say stalkers and criminals are using similar technologies to track cellular communications. While EFF attorney Fakhoury says, "It's still too early to tell how [widely deployed] Stingrays will be," he also notes that "it's troublesome to have devices like this that can look around everywhere and get the information of every phone in an area." The EFF calls Stingrays the "biggest technological threat to cellphone privacy you don't know about."[50]

Stingrays will pose an even greater menace to privacy if they are paired with the unmanned aerial vehicles (UAV) or unmanned aircraft systems (UAS) known as drones. Domestically drones are used for law enforcement, firefighting, search and rescue efforts, disaster relief, and weather monitoring, while the military uses them abroad for reconnaissance and surveillance missions and attacks against alleged terrrorists. Drones aren't focused on capturing smartphone data, but privacy groups say the vehicles can be outfitted with devices that intercept phone calls and text messages. Independent security researchers already have developed a small drone that can spy on data that smartphones send over Wi-Fi, such as e-mail usernames and passwords and credit card information.[51] "At some point drones and Stingrays will converge," predicts Fakhoury.

Though drone privacy regulations have yet to be implemented on a federal level, law enforcement agencies have been flying the vehicles throughout the country for several years. U.S. Customs and Border Protection (CBP) owns approximately ten drones,[52] which it deploys along the U.S.-Mexico and U.S.-Canada borders to "identify and intercept potential terrorists and illegal cross-border activity."[53] CBP also flies its drones on behalf of other government agencies, including the FBI, U.S. Immigration and Customs Enforcement (ICE), the Drug Enforcement Administration (DEA), the U.S. Marshals Service, and county sheriff's departments, and it did so nearly 700 times between 2010 and 2012.[54]

The drone privacy issue is expected to take on more urgency. In March 2014, a federal judge ruled that the de facto ban the Federal Aviation Administration (FAA) had placed on commercial drones in U.S. airspace was not legally binding. The decision would have allowed photographers, filmmakers, surveyors, and news organizations to deploy unmanned vehicles over the continental United States, but was stayed because the FAA quickly appealed the decision. The resulting ruling is still pending, but privacy groups worry that when commercial drones are integrated into the skies, some of them will be used for surveillance, due to their low operational costs and advanced technological capabilities.

PUBLIC RESPONSE

Not everyone finds these emerging technologies intimidating. People's notions of privacy are highly individual and often vary according to age. In an April 2013 survey of 3,900 consumers in 13 countries, nearly two out of three (65 percent) 18- to 34-year-olds described themselves as uninterested in privacy matters related to online and mobile communications.[55] The next age group, of 35- to 44-year-olds, expressed only slightly more interest in guarding their digital privacy.

Yet polls that asked Americans specifically about warrantless phone searches, location tracking for marketing purposes, and NSA data collection found the majority of respondents object to these actions. More than three out of four respondents to a 2012 University of California–Berkeley study said police should get court permission before searching a phone during an arrest.[56] In the same study, 70 percent of respondents said they would "definitely not allow" carriers to use their locations to tailor ads to them.[57]

Sentiment about NSA data tracking was also negative. In a November 2013 *Washington Post* poll 69 percent of respondents said they were concerned about carriers' collection of their personal information, and 66 percent said they were concerned about the NSA's collection and use of their personal information. A January 2014 poll from the Pew Research Center and *USA Today* found the majority (53 percent) of respondents opposed the NSA's bulk-phone-records program. The results represented a notable increase in negativity from July 2013, when a majority (50 percent) of respondents said they approved of the program and only 44 percent disapproved.

Young people are not always indifferent to privacy concerns, either. In an August 2013 Pew study, 59 percent of teenage girls said they had disabled location tracking on their smartphone apps due to privacy worries. (Teenage boys expressed less concern, with 37 percent of them taking this precaution.) A large majority of young Americans are also clearly against government collection of their phone and online communications for NSA-like purposes. An October-November 2013 Harvard University Institute of Politics survey of 18- to 29-year-olds found that only 14 percent of respondents were in favor of the government tracking their phone calls and GPS location, even if doing so would "aid national security efforts."[58]

It's clear the law needs to catch up with technology. The mass aggregation and analysis of information enabled by smartphones, affordable data storage, cloud computing, and supercomputers would have been unfathomable in the 1970s and 1980s, but that is when many of our current electronic privacy laws were written or legal precedents were established.

Much of the government's legal rationale for arguing that people can't expect privacy regarding their phone records stems from a 1979 legal case called *Smith v. Maryland*, which involved a phone company installing a device called a "pen register" on a robbery suspect's home phone line in response to a police request (without a warrant). The device recorded the numbers the suspect—Smith—dialed and soon showed an outbound call to the house of a woman whose purse had been stolen, confirming him as a suspect in the robbery. Smith was convicted and argued that the warrantless use of a pen register constituted a search that invaded his privacy, but the Supreme Court rejected his claim, writing in its opinion: "All [telephone] subscribers realize . . . that [their] phone company has facilities for making permanent records of the numbers they dial."[59]

Nearly 35 years after *Smith v. Maryland*, government lawyers and officials cited the case to defend the NSA's phone metadata program, even though the NSA's program involved millions of Americans rather than one robbery suspect and the information that police learned about Smith from looking at his phone records for a few days was far more limited than what law enforcement could deduce from analyzing years of people's cellphone and smartphone records today.

Stingrays and drones also gather much more information than would have been anticipated in the 1970s and 1980s, but experts say courts will probably draw on decades-old rulings to make decisions about those technologies, too. Says Fakhoury: "There was a Supreme Court case in the 1980s where the police department took a plane and flew it to look in a guy's backyard to see if he was growing marijuana. The Court said that was OK and didn't violate the Fourth Amendment. Under that precedent, a fight against drone [spying] will be tough."

Lawmakers have introduced bills that aim to amend such outdated digital privacy laws as the 1986 Electronic Communications Privacy Act (ECPA), which includes the Stored Communications Act. When these laws are finally revamped, privacy groups want them to mandate search warrants for all forms of electronic communications, including content, metadata, and location data. "Smartphones are becoming more accurate and precise," says Fakhoury. "If we want to future-proof these laws so they can withstand the next thirty to forty years, we need to stop thinking about everything as content or noncontent and start considering what different types of data reveal about a person."

8

Looking Toward the Future

According to a 2013 Pew survey, 15 percent of U.S. adults don't use the Internet at all. For many adults, staying offline is a deliberate decision, but about one in five (19 percent) cited price and 7 percent cited access challenges to explain why they do not go online.

Some groups contend that smartphones can be an effective way to narrow the digital divide in the future. They note that minorities are not only more likely to own smartphones than white Americans, they also tend to do a broader range of activities on them compared to white Americans. Pew's data also supports these findings: 64 percent of blacks and 60 percent of Hispanics owned smartphones in 2013 compared to 53 percent of whites, and blacks and Hispanics were more likely than whites to use their phones to access the Internet, send or receive e-mail, and download apps. Youth is a factor here. The white American population is aging, while the number of young people from minority backgrounds only continues to grow, and those demographic trends influence smartphone ownership rates. Pew surveys also show that young adults, nonwhites, and the less affluent are more likely to use their phones as their primary source of Internet access. These groups often can't afford laptops and home broadband connections, but they can and do purchase smartphones, frequently on lower-cost, prepaid plans.

The Minority Media and Telecommunications Council (MMTC), a nonprofit advocate for diversity and civil rights in the media and telecom industries, has argued that wireless broadband, as accessed on a smartphone, "offer[s] all or nearly all of the benefits of fast home broadband service but at a far lower price point—thus enabling adoption."[1] The cellphone–smartphone industry takes a similar stance.

CTIA, the wireless industry association, has called mobile broad-band a "great equalizer" that helps consumers and drives the U.S. economy.[2]

HOW AGE AND INCOME AFFECT SMARTPHONE OWNERSHIP

In the United States, youth and affluence correlate strongly with smart-phone ownership.

Age group	The percentage that owns smartphones
18–29	83%
30–49	74%
50–64	49%
65+	19%

Household income	The percentage that owns smartphones
Less than $30,000/yr	47%
$30,000–$49,999	53%
$50,000–$74,999	61%
$75,000+	81%

Education attainment	The percentage that owns smartphones
High school grad or less	44%
Some college	67%
College+	71%

Residential area	The percentage that owns smartphones
Urban	64%
Suburban	60%
Rural	43%

Source: Pew Research Center Internet Project Survey, January 2014

Others note that smartphones help close the broadband adop-tion gap but have limits when used as a person's sole connection to the Internet. While Pew has acknowledged that smartphones consti-tute "an alternate form of home Internet access,"[3] the think tank has questioned at the same time, whether smartphones offer the same economic benefits as traditional broadband access. "It is unclear whether 3G or 4G smartphones qualify as 'broadband' speed, or if smartphones can otherwise offer the same utility to users as a dedicated high-speed home internet connection," wrote Pew researchers in 2013.[4] In the same year, a paper published by the nonprofit think tank New America Foundation similarly posited that mobile Internet access was not comparable to PC-based Internet access. The paper noted: smartphones have less memory, speed, and storage capacity than computers; they sometimes cannot access all Internet sites; and they are less conducive to creating content than enabling the user to consume it. Because of these "potentially significant compromises and shortcomings," the paper's conclusion cautions against policies that "emphasize mobile access and largely abandon any emphasis on PC-based access."[5]

Susan Crawford is one of the most prominent voices on this side of the debate. Crawford is a professor, author, and telecommunications policy expert who spent 2009 as the special assistant to President Obama for science, technology, and innovation policy; she resigned after eight months to return to academia. She has pointed out the difficulties of doing work on smartphones, with their relatively small screens and keyboards, and over wireless connections, which are slower than wire-line broadband and subject to carrier data caps. In a 2011 commentary published in the *New York Times*, Crawford wrote, "It is hard to type up a résumé on a hand-held device; it is hard to get a college degree from a remote location using wireless. Few people would start a business using only a wireless connection."[6] Crawford contends that wireless connections are "second-class" and a mere "bike path" compared to the "superhighway" of high-speed wire-line connections.[7] She has warned that the split between people who can afford wire-line, in-home broadband and those who make do with phones relegates "racial minorities and poorer and rural Americans" to "lower-quality health services, career opportunities, education and entertainment options."[8]

Benton Foundation policy director Amina Fazlullah largely agrees. She says cellphones and smartphones are indispensable for connecting low-income families and underprivileged minorities to doctors, emergency services, and educational and job resources. At the same time, she says, mobile devices offer only a "second tier" of services in terms of Internet access and the ability to produce content, and thus "inadvertently turn vulnerable populations into one-way [Internet] users" who mostly consume rather than produce content. She believes this holds true even though smartphones are getting bigger and faster, because mobile data plans continue to be limited (or else pricy or slow), and smartphone screens and keyboards are still small relative to computers. Fazlullah says, "You can't spend a lot of time researching on a phone. It's hard to develop your own website or app on a phone screen. Watching Khan Academy [e-learning] videos for tutoring or watching job-training videos or making Skype calls with distant family members are legitimate [Internet] uses that low-income Americans should have access to—and all take up a significant amount of bandwidth."

Fazlullah and Crawford consider smartphones and mobile broadband to be a short-term solution or a first step to bridging the digital divide that needs to be followed by a long-term solution. Fazlullah says the government should increase its support for "real, robust" broadband, not just mobile broadband.

The Obama administration has been a vocal supporter of mobile broadband, whether for smartphones or laptops. In 2011, during his annual State of the Union address, Obama announced a national wireless initiative to bring 4G connectivity to at least 98 percent of Americans within five years. The initiative proposed holding government auctions of wireless spectrum and earmarking $5 billion of the proceeds to building 4G networks in rural areas and $3 billion to R&D for emerging wireless technologies. The White House's announcement said the initiative would enable the United States to "win the future by catalyzing the build-out of high-speed wireless services that will enable businesses to grow faster, students to learn more, and public safety officials to access state-of-the-art, secure, nationwide, and interoperable mobile communications."[9]

Most of these aims have yet to be met, but the FCC did establish what it called a "mobility fund" in 2012 as part of a broader Connect America Fund (CAF), replacing a portion of the Universal Service Fund (USF), which was established in 1997 to subsidize telecommunications services for underserved groups, including rural and low-income Americans, schools, and libraries. By 2017, the part of the USF that used to underwrite landline phone service in rural areas will be fully transferred to the CAF to underwrite wire-line and wireless broadband service in those areas. The FCC is also modernizing another USF program, known as E-rate, directing those funds to deploying high-speed broadband (wire-line and wireless) in schools and libraries. The commission has already committed $2 billion to this ConnectED initiative and several large tech companies, including AT&T, Sprint, and Verizon, donated more than $750 million more in gadgets, software, and free wireless service.

Fazlullah says the Benton Foundation lauds the Obama administration's efforts to expand spectrum and mobile broadband availability. But she also says the FCC and White House should take care to listen not just to the technology industry, which has enormous power over policy makers and can overwhelm public interest advocates and the voices of the public. Instead, the government's ultimate objective should be to give all Americans "strong and fair access" to high-speed Internet. "If mobile broadband is a pathway to making sure everyone has broadband they can use for laptops and computers, that's fine," Fazlullah says. "But if it means everyone just ends up with a phone in their hands, that's not going far enough." National policies that focus on improving mobile broadband can make it harder to push for higher-quality connectivity, because lawmakers may assume a certain level is sufficient and stop there. "When vulnerable populations get minimal access to broadband, it can take the wind out of the sails for advocates in DC and elsewhere [who are pressing for better broadband access]," says Fazlullah.

LIFELINE

Since cost is a major barrier for low-income Americans who have yet to adopt smartphones, it looks likely that the government will start subsidizing smartphones and smartphone service through Lifeline, a federal program that provides landline and wireless service discounts. Lifeline was established in 1984 and implemented in 1985, under the

Reagan administration, as a landline-only program. At the time the FCC, which oversees the program, said, "Access to telephone service has become crucial to full participation in our society and economy." [10] The George W. Bush administration extended Lifeline to include prepaid wireless service, initially in 2005 for people displaced by Hurricane Katrina, and later in 2008 for all Americans.

Approximately 13.4 million low-income Americans currently receive discounts through Lifeline, which is supposed to be limited to one phone or one account per household. People can qualify for it by having a household income at or below 135 percent of the federal poverty guidelines, which in most states is about $32,000 a year for a family of four, or by participating in a federal assistance program, such as food stamps, Medicaid, or Section 8 public housing.

Other Americans indirectly help pay for Lifeline. The program is technically supported by carrier contributions to the USF, but carriers pass this expense on to their subscribers by charging them small fees (around $2.50) on their monthly bills. A nonprofit called the Universal Service Administrative Company (USAC) administers the fund, which it uses to reimburse carriers that provide cellphone service to Lifeline recipients. The carriers receive about $9 per month per subscriber. Carriers can also earn more if they serve subscribers in states that have their own, supplemental Lifeline programs. The two largest Lifeline providers (out of about 2,000 companies nationally) are Miami-based SafeLink Wireless and New Jersey–based Assurance Wireless. Assurance is a subsidiary of Virgin Mobile, which is owned by Sprint, and SafeLink is a subsidiary of TracFone Wireless, which itself is a subsidiary of the Mexico-based carrier América Móvil, one of the world's largest multinational carriers.

Lifeline subsidizes phone service, not phones. But carriers that receive Lifeline funding often give subscribers free cellphones to entice them into signing up for the service. The phones are usually basic models and generally include 250 prepaid, complementary voice minutes a month, and sometimes text messages. These phone handouts have led critics to call Lifeline the "Obama phone" program. Some people mistakenly believe that the Obama administration mandated the free phones to curry support and votes in the 2012 presidential election—a false notion that was reinforced by a much-discussed September 2012 video in which an Obama supporter in Ohio urged passersby to reelect Obama, and said, "He gave us a phone, he's gonna do more!" [11]

The FCC is considering expanding Lifeline to add smartphones as an official option, which could be controversial. The program already has plenty of critics who say it is wasteful. Between 2008 and 2012, Lifeline expenses nearly tripled, to $2.2 billion, due to increased enrollment, and a portion of that money fraudulently enriched some companies. For example, a 2013 Indiana Utility Regulatory Commission investigation found that Lifeline provider TerraCom approved cellphone applications bearing signatures that didn't match the names on the forms and listed addresses that were actually vacant or abandoned homes. A previous investigation by the Scripps News service revealed that sales agents for TerraCom and its sister company YourTel America had completely forged Lifeline applications in several states to maximize revenues and increase their sales commissions.

Dozens of Congress members have pressed for Lifeline reforms in recent years. The FCC increased oversight of the program in 2012 and says it has saved hundreds of millions of dollars by requiring carriers to: verify subscribers' eligibility before activating the service; recertify subscribers each year; and de-enroll those who stop using their phones for more than two months. In 2013, the agency also fined 11 Lifeline providers $90 million for fraudulent reimbursement requests, and it began collecting Lifeline subscriber enrollment information in a central database—a long-awaited and requested reform designed to reduce duplicative accounts.

Fazlullah says Lifeline had "some legitimate issues," mostly stemming from the service providers, and she cites the FCC clampdown on corrupt carriers as evidence that the program has been cleaned up in the past two years. She adds, "I'm relatively confident the FCC is on the right track, and that if they find more fraud, waste, and abuse [in Lifeline], they will write new rules or clarify the rules or push harder on the service providers."

THE NEXT BILLION

Reducing smartphone costs is an important goal for the smartphone industry, too, if for less altruistic reasons. Analysts say smartphones are starting to saturate developed markets, and sales of high-end phones have slowed. To keep flourishing, the industry must figure out how to make phones affordable. For the next few years, smartphone shipments are expected to increase at least twice as fast in emerging markets as in developed ones—more than 23 percent annually in the Asia Pacific and Latin American regions versus 8 percent in North America and 11 percent in Europe.

TOP 20 SMARTPHONE COUNTRIES

Smartphone Ownership by Percentage of Population

Rank	Country	Smartphone Penetration
1	United Arab Emirates	73.8%
2	South Korea	73%
3	Saudi Arabia	72.8%
4	Singapore	71.7%
5	Norway	67.5%
6	Australia	64.6%
7	Sweden	62.9%
8	Hong Kong	62.8%
9	United Kingdom	62.2%
10	Denmark	59%
11	Ireland	57%
12	Israel	56.6%
13	United States	56.4%
14	Canada	56.4%
15	Spain	55.4%
16	Switzerland	54%
17	New Zealand	53.6%
18	Netherlands	52%
19	Taiwan	50.8%
20	Austria	48%

Source: Google's OurMobilePlanet.com, data collected via online questionnaire, January-February 2013

Africa, Asia, eastern Europe, and South America are the new smartphone growth engines. The market research firm IDC expects Indian and Brazilian purchases to surge, switching the world's top-five smartphone markets from China, the United States, Britain, Japan, and Brazil in 2013 to China, the United States, India, Brazil, and Britain by 2017. Most of those new consumers, especially in China, India, and Brazil, require low-cost smartphones.

Phone makers, carriers, and platform providers have anticipated this shift. For the past few years they have talked up the importance of connecting the next billion users to smartphones and fast mobile data networks. Expanding smartphone and mobile Internet access is a business goal that has the bonus effect of helping less affluent people. The World Bank estimates that a 10 percent uptick in broadband can increase a developing country's GDP by 1.4 percent. In much of the developing world, broadband is being deployed as wireless broadband, because wireless networks are cheaper to build than wire-line networks. The spread of wireless broadband, in turn, helps sell smartphones by enabling people to fully realize their potential as Internet-connected handheld computers.

The industry is still debating how far prices have to drop to enable another billion people to buy smartphones and get online. Carriers do not subsidize phones in the majority of emerging markets, so consumers must pay a phone's full cost upfront. A recent report from mobile marketer Upstream, which surveyed consumers in Brazil, India, Nigeria, and Saudi Arabia about their mobile preferences, found almost half of respondents were willing to spend $100 to $300 on a smartphone, but nearly a third (29 percent) didn't want to spend more than $100. Mechael of the mHealth Alliance says international development workers typically cite $50 as the price that would enable people in underdeveloped economies to obtain smartphones. Manoj Kohli, the former CEO of Bharti Airtel, India's largest wireless carrier, has challenged manufacturers to produce $30 smartphones.

Until recently, a reliable $30 smartphone seemed years away. But in February 2014, Mozilla announced a $25 Firefox OS reference design, created in partnership with the Chinese chip maker Spreadtrum, which should make a $30 smartphone possible. Mozilla has targeted emerging markets from Firefox OS's earliest days, stating in its 2013 launch announcement that it wanted the operating system to "bring the freedom and unbounded innovation of the open Web to mobile users everywhere" and "meet the diverse needs of the next two billion people online."[12] In line with Mozilla's populist mission, early Firefox OS phones cost about $100 unsubsidized and were targeted at first-time smartphone buyers in Brazil, Colombia, Mexico, Peru, Uruguay, and Venezuela, among other countries. Though Firefox OS ended its first year of availability with less than 1 percent of the global smartphone market, the platform saw a lot of activity, with four carriers launching a total of three Firefox OS phones in 15 European and Latin American countries. More phones and handset partners are on the way.

Tier-one smartphone makers aren't likely to produce one of Mozilla's $25 smartphones, but they have been giving their ultra-budget phones some smartphone zing. Nokia is a good example. The cheapest phone in its Windows Phone–powered "Lumia" series, the Lumia 630, is priced at around $160. In 2011, Nokia launched a line of pseudo-smartphones called Asha that typically sell for less than $100 and are popular in India and Southeast Asia. Asha phones are aimed at first-time mobile Web users and sport touchscreens and Web browsers that compress content to reduce mobile data costs; they can download apps from a small, Asha-specific app store.

Nokia continues to introduce cheaper Asha phones, with the cheapest, the Asha 230, costing just $62. Nonetheless, low-cost Android phones outsell Ashas in many markets, because they have more apps, and in February 2014 Nokia introduced an affordable family of smartphones based on Android open-source code. Nokia says the phones, which have names beginning with Nokia X, "combine Nokia design, build quality, and services with the ability to run Android apps."[13] At $122 to $150 each, they are priced between Nokia's Asha and Lumia lines, and are intended to "appeal to new smartphone users" in China, India, and Russia, who are "looking for popular apps and their first cloud services."[14] This last part is key; though Nokia is using Android for its X smartphones, it is shipping them with Nokia and Microsoft services, including Nokia maps, Skype, Bing, and Microsoft's cloud storage and free Outlook.com e-mail service. In part this is because

Google did not approve the phones and thus won't let Nokia preload popular Android services on them, but it's also because Nokia is using the phones to "push the masses towards Microsoft's ecosystem," as The Verge put it.[15] Or as TechCrunch said, "[Nokia X] is one way for Nokia to fight Android's ecosystem dominance—by piggybacking on it."[16]

[Top] Nokia Asha 230, Nokia's $62 pseudo-
smartphone (*Sam Churchill/Flickr*)
[Bottom] Nokia X, Nokia's Android-based,
entry-level smartphone for emerging markets
(*Mac Morrison/Flickr*)

Samsung is one of the Android phone makers that drove Nokia to finally adopt—though not embrace—Android. Since February 2013 Samsung has sold a line of entry-level app- and Internet-enabled phones called REX that are a stepping-stone between basic handsets and Android phones. REX phones are comparable to Nokia's Asha handsets in terms of price ($70 to $110), features (touchscreens, select apps, and streamlined browsers), and target markets (India, Asia, and South America). REX is supposed to help people gradually step up to Samsung's more expensive Android smartphones, but Samsung has found more success in emerging markets with its low-cost Galaxy phones, some of which are priced close to $100. Tricia Wang, a cultural sociologist who studies mobile computing's impact on "nonelite communities," says migrant workers will save up for months to buy smartphones, in part because smartphone ads in developing countries "plant ideas and dreams and images into people's heads of what they can aspire to." The aspirational nature of these ads and the smartphones that they feature have helped Samsung sell more low-cost smartphones worldwide than its rivals, including Nokia.

In a sign of their commitment to emerging markets, both Samsung and Nokia are founding members of Internet.org, a Facebook-led partnership to make the Internet affordable in the developing world. The initiative was announced in August 2013, and one of its three main goals involves creating "lower-cost, higher-quality"[17] smartphones so that underserved communities can get online. The group's other two goals involve developing technologies to make mobile data networks run more efficiently and devising business models that support the Internet.org mission. In February 2014, Internet.org said Nokia would provide some of its smartphones at a subsidized price for a pilot project that connects Rwandan students to online educational content.

Among smartphone platform providers, Google is the most enthusiastic about tapping emerging markets. Reaching the next billion has been a Google theme for years. In 2011, executive chairman Eric Schmidt said increasing the availability of affordable smartphones "in the poorest parts of the world" was one of Google's three biggest strategic initiatives.[18] "We envision literally a billion people getting inexpensive, browser-based touchscreen phones over the next few years," he wrote in a *Harvard Business Review* blog post.[19] During a 2013 AllThingsD conference, Schmidt reiterated that Android's goal is "to reach everyone. . . . The way that's going to happen is with the debut of low-end devices from manufacturers, primarily in Asia."[20]

Google has already succeeded in making Android the developing world's leading smartphone platform. In China and India, Android comprises as much as 90 percent of smartphone shipments. In fact, Android is so dominant in China that the country's Ministry of Industry and Information Technology published a white paper in 2013 warning that domestic smartphone makers were becoming "too dependent" on the platform.[21] Nonetheless, Google actually derives little profit from China-based Android devices, because many Chinese smartphone makers swap Google's search engine and apps for locally developed services. The mobile metrics firm Distimo says only 3.5 percent of Chinese Android devices have Google Play installed.

While Google benefits less from these unofficial Android phones, they certainly help boost its market share. In 2013, Google said Android was growing three times more quickly in developing countries than in developed ones. Google fosters this growth in a few ways. It lets smartphone makers use Android without licensing fees, which helps them manage their expenses. It also tweaked Android so the operating system runs smoothly on lower-end phones.

From 2013 through January 2014, Motorola was a key part of Google's affordability campaign. In November 2013, Motorola unveiled the first low-cost smartphone it designed with Google's input and aimed the device at developing markets, including Brazil and India, and the U.S. prepaid market. The $179 Moto G resembles Motorola's Moto X flagship, but to reduce costs, the Moto G has a smaller display, half the megapixels (5 versus 10), half the onboard storage (8 GB versus 16 GB), and no 4G LTE connectivity.

Apple and Microsoft have their own strategies for tapping emerging markets. Unlike Google they are intent on keeping their user experience consistent and, in Apple's case, protecting high smartphone margins. Many analysts and investors want Apple to introduce a truly low-priced iPhone—lower than the iPhone 5C—but Tim Cook has said he is not interested in that part of the smartphone market. In a 2013

Bloomberg Businessweek interview, Cook declared, "We're not in the junk business, . . . because it's just not who we are."[22] Instead Apple is expanding its retail presence in several developing countries, opening its first Latin American store in Brazil's Rio de Janeiro in February 2014.

Microsoft didn't appear to have a strong emerging-markets game plan until 2014. But that year, it moved quickly to lower the costs and reduce the work associated with producing Windows Phones. Within the space of about six weeks, it launched an explanatory website for new Windows Phone makers, introduced smartphone reference designs that used lower-end Qualcomm processors, and made its operating system free. The moves were aimed at expanding Windows Phone's reach globally, but particularly in emerging markets. Around the same time, Microsoft announced a slew of new and returning Windows Phone manufacturing partners, many of them based in China or India. A few were large companies, such as Lenovo and LG, but most were lesser-known, Asia-only manufacturers. To get these latter companies on board, Microsoft had to waive its licensing fees. As a "senior executive" at an unnamed Indian phone maker told the *Times of India*, "Windows Phone still doesn't have [a] lot of appeal in the market, but now that it doesn't have any license fee, it becomes easier for us to experiment with it."[23]

Microsoft has also said it will use Nokia's cheaper non-smartphones as "an on-ramp to Windows Phone"[24] to "[continue] to connect the next billion people."[25] And, in Africa specifically, Microsoft has a $70 million 4Afrika initiative that leverages Microsoft technology and resources to "improve [Africa's] global competitiveness."[26] One of the initiative's goals is to help place tens of millions of smart devices, including smartphones, "in the hands of young Africans"[27] by 2016. As part of the program, Microsoft and its device partners are producing low-cost, Africa-centric, 4Afrika-branded Windows Phones. The first 4Afrika smartphone launched in 2013 in seven African countries for around $150 unsubsidized, about the same price as Nokia's budget Lumia phone, and it was preloaded with region-specific apps, such as an app about Nigeria's homegrown film industry, Nollywood. It was a great price for a quality phone but nearly twice as expensive as entry-level Android African ones.

BlackBerrys might seem a strange fit for developing markets, but their generally reliable, low-bandwidth network connections and free BBM messaging service have kept them popular in countries such as Indonesia and Nigeria. BlackBerry is now counting on these consumers to help keep it afloat while it reworks its enterprise strategy. The first BlackBerry-Foxconn phone to target this audience is the Z3, a touchscreen phone that launched in Indonesia in May 2014 for under $200, unsubsidized.

THE POWER OF CHINDIA

The fiercest smartphone wars in the developing world are being fought in China and India. China's continued rollout of 4G LTE service, which the government officially approved in December 2013, will further expand its lead as the world's largest smartphone market. Established smartphone makers such as Samsung not only are battling to lead the Chinese market, they are also trying to fend off China-based upstarts that are encroaching on their global sales. Telecommunications infrastructure equipment maker Huawei and Lenovo are two of China's top contenders. According to IDC, they currently rank number three and number four, respectively, in global smartphone shipments—behind Samsung and Apple and ahead of Korea's LG. Strategy Analytics says both Huawei and Lenovo grew their smartphone shipments about two times faster than the global industry average in 2013.

Huawei and Lenovo's rapid rise reflects China's growing prominence in the industry. Neither company produced high-end smartphones for international consumers before 2011. Instead Huawei and Lenovo focused mostly on China, first with basic phones and later with Android smartphones, starting with the U8230 in 2009 (Huawei) and the LePhone (Lenovo) in 2010. Then they leveraged their China experience and profits to expand internationally and produced premium Android smartphones.

Chinese smartphone makers enjoy several advantages over Western, Korean, and Taiwanese companies. They possess a competitive edge in their enormous home market, where they each sell millions of smartphones a year. They also tend to have lean operations and are located close to their manufacturing facilities, enabling them to move quickly and charge less than their competitors. Market trends also favor Chinese smartphone makers. As Kim Yoo-chul of the *Korea Times* notes, Huawei and Lenovo phones appeal to the growing ranks of consumers who prefer (or just need) smartphones that have competitive hardware specifications but affordable prices.

What Chinese smartphone companies lack are international brands. A recent marketing survey that asked Americans to name a Chinese brand found that a whopping 94 percent of respondents couldn't think of a single one. Just 2.5 percent of respondents mentioned Lenovo, though it has sold computers in the United States since 2006, a year after it completed its acquisition of IBM's PC business. And only 1.1 percent of respondents named Huawei, though it ranks number 315 on *Fortune*'s list of the 500 largest global companies, more than 10 notches above American Express and more than 20 above Air France–KLM. Kim says Huawei and Lenovo "suffer from a lack of brand power and [distinctive] product design, [but] . . . will try hard to address such issues," given their "strong ambitions to gain scale in handsets."

Huawei and Lenovo have been building up their brands. In 2012 Huawei established a mobile device design office in London to better target the European market, and in 2013 the company touted its Ascend P2 as the world's fastest 4G smartphone and its Ascend P6 as the world's thinnest smartphone. That year Wan Biao, who was then CEO of Huawei's mobile devices division, told the British *Telegraph*, "If you look back five years ago, Apple is small, Samsung is not so big. You can't see where we'll be in five years. At least top three [in the world]. Maybe number one."[28] Brand recognition is one of the primary reasons Lenovo bid for Motorola. In a press release announcing the deal, Lenovo CEO Yang Yuanqing called Motorola an "iconic brand" that, along with Motorola's people and product resources, will "immediately make Lenovo a strong global competitor in smartphones."[29] Lenovo also pursued a bid for BlackBerry in 2013, according to the *Globe and Mail*, but backed off when the Canadian government expressed national security concerns.[30]

Chinese companies Coolpad, TCL, and ZTE are now attempting

to follow Huawei and Lenovo into international markets. Coolpad/ Yulong is China's third-largest smartphone maker, and it ventured into the United States in 2012 with one smartphone at MetroPCS, a smaller U.S. carrier now majority-owned by T-Mobile. TCL sells a range of consumer electronics and appliances in China, but to increase its brand recognition it recently purchased naming rights to Hollywood's famed Grauman's Chinese Theatre, changing it to TCL Chinese Theatre. Outside China TCL sells smartphones under the brand Alcatel, and it bought product placement for one of its flagship smartphones, the Alcatel One Touch Idol, in the 2013 movie *Iron Man 3* to promote that brand in the global market. T-Mobile and MetroPCS currently offer Alcatel smartphones. ZTE, like Huawei, makes telecom infrastructure equipment and has been trying to extend its brand and sales outside China for years. In 2013, ZTE became a sponsor of the Houston Rockets basketball team and said it plans to be one of the world's top three cellphone makers by 2016. Sprint currently sells a ZTE smartphone.

The Chinese start-up Xiaomi is also coming on strong. It didn't start making smartphones until 2011 but already sells 19 million units a year in China, Hong Kong, and Taiwan, recently expanded distribution to Singapore, and plans to enter Brazil, India, Malaysia, and Russia. Consumers scoop up Xiaomi phones because they have the same high-end features as premium smartphones but cost less than half as much, with the most expensive Xiaomi model selling for around $320, unsubsidized. "We essentially price our phones at bill-of-materials [cost]," explained Xiaomi president Bin Lin in a 2013 AllThingsD interview.[31] Xiaomi, which describes itself as "a mobile internet company" on its website, has said its phones give it a platform for selling software and services, including games and mobile messaging, much the way the Kindle e-readers spur book sales for Amazon. The start-up shows so much promise that investors valued it at a whopping $10 billion in its latest funding round.

In fact, Xiaomi now sells more smartphones in China than Apple, according to the market research firm Canalys. The iPhone's high price holds Apple back in China; it ranks number six in smartphone sales behind Samsung, Lenovo, Coolpad, Huawei, and Xiaomi. As the market researcher Kantar Worldpanel ComTech has noted, Chinese consumers "are increasingly opting for a high-spec local brand over a low-spec global equivalent,"[32] so they can get the best value for their hard-earned money. Chinese carriers subsidize the iPhone, but their prices are still higher than those of other smartphones; the unsubsidized price of the iPhone 5S is more than $850. (The older iPhone 4S, which Apple also sells in China, is much cheaper, at $528.)

IPHONE 5S GLOBAL PRICE INDEX

Where the iPhone is cheapest and most expensive, based on the unsubsidized price of an iPhone 5S with 16GB of memory

Country	Price with sales tax	Country	Price as percentage of GDP per capita
United States	$707.41	United States	1.3682%
Hong Kong	$720.62	Hong Kong	1.4148%
Taiwan	$760.30	Canada	1.8331%
Canada	$775.72	Taiwan	1.9822%
South Korea	$831.46	United Kingdom	2.4527%
China	$868.14	Germany	2.4756%
India	$872.29	South Korea	2.6024%
United Kingdom	$896.94	Finland	2.6759%
Russia	$914.61	France	2.7508%
Finland	$957.21	Spain	3.1845%
Germany	$957.21	Italy	3.3486%
Spain	$957.21	Russia	5.2210%
France	$970.90	China	9.5874%
Italy	$998.29	Brazil	10.1849%
Brazil	$1,196.42	India	22.6981%

Source: Mobile Unlocked; data current as of November 2013

In recent years Apple has made considerable effort to boost its China business, signing contracts with two of China's largest carriers in 2012 (one of which was an extension of a 2009 contract) and announcing plans in April 2013 to double its China region stores within two years. Apple also struck an iPhone deal with China Mobile, the world's largest carrier with an estimated 760 million subscribers, in December 2013 following years of negotiations. In a CNBC interview prior to the China Mobile iPhone launch, Tim Cook called the occasion a "watershed day."[33]

The burgeoning power of Chinese consumers and smartphone makers raises the prospect that China will eventually influence the types of smartphones the rest of the world will use. In another sign that the East is coming west in the smartphone industry, Japanese app developers are increasingly generating global hits. Puzzle & Dragons, a role-playing puzzle game from Japan's GungHo Online Entertainment, was the highest-grossing game in the App Store and Google Play for all of 2013, according to App Annie, the mobile app analytics firm. Brave Frontier, a gods-and-beasts role-playing game from gumi Inc., a Japan-based gaming company, was one of the top 25 highest-grossing games in the U.S. App Store and Google Play in the spring of 2014. "More and more globally successful games are coming from Asia," says Ollie Lo, App Annie's vice president of marketing. "The Asian guys are absolutely killing it, and not just in their own countries."

Lo says Apple and Google's centralized app stores make it easy for app developers to go global. Japanese developers also have more experience crafting mobile games than their Western counterparts. "Asian companies have been monetizing on mobile for a decade," says Lo. "That know-how is coming through." Chinese developers don't have as much experience, but they are learning quickly. Apple has said it has more registered iOS developers in China than in the United States—more than 500,000 versus 300,000.

Western developers are trying to move in the other direction but are having less success. Besides needing to translate their apps into Asian languages, they have to strike up partnerships with Asian social networks and payment providers. "A lot of Western [app] publishers find it hard to enter Asian markets," says Lo. The *Wall Street Journal* recently described the Chinese app market as "untapped and untamed," because it offers potentially huge business opportunities but also greater risks, including widespread app piracy and the need to work with dozens of local Chinese app stores.[34]

India's smartphone market is even more untapped than China's. Like China, India has an enormous population and a burgeoning middle class. India is the world's second-largest market for cellphones, but its retail distribution is fragmented between many different companies, and its 3G coverage is spotty outside of cities. Just 6 percent of Indians own smartphones, compared to 35 percent of Chinese people, based on Canalys data. Smartphone companies are sparring over the opportunity to sell hundreds of millions of Indians their first smartphones, and their rivalry has turned into a war of escalating consumer incentives.

Samsung leads the Indian smartphone market, due to its wide array of phones and affordable prices, around $100. Sony ranks number four in India, and Apple is number six, because of the iPhone's high price. Since most Indian carriers don't subsidize phones, iPhones can cost more than $850 in India, just like in China. Apple also has no official stores in India and didn't launch iTunes there until 2012. To compensate for these limitations, Apple introduced a range of offers for Indian iPhone buyers in 2013, including: discounts for students and people who traded in competing smartphones; monthly installment payment plans; and "cash back" rebates for iPhones purchased with credit cards. The promotions boosted iPhone sales by as much as 300 percent in just a few months. But Apple's success also spurred rivals, such as Samsung, to establish or increase their own smartphone offers, which ultimately limited Apple's gains. In January 2014, Apple tried reclaiming market share by bringing back the iPhone 4 at its lowest price ever—$365, unsubsidized. The move, which was specific to emerging markets, came several months after Apple had discontinued the iPhone 4 globally.

Like China, India has a crop of local smartphone makers that specialize in low-priced Android phones. Micromax is India's largest domestic smartphone maker, and it sells about two-thirds as many units in India as Samsung. Micromax's flagship phone sports a five-inch HD display, a 16-megapixel camera, and some innovative sensing technologies. It costs half the price of the Galaxy S5 and iPhone 5S. Karbonn is India's second-largest local smartphone maker. It was established a year after Micromax started making cellphones, in 2009, and offers similar phones for similar prices. Neither company has yet made a global impact, but a 2014 *New York Times* article quoted Rahul Sharma, Micromax's cofounder, as saying, "We are going up notch by notch. We are changing the tonality and cool factor of the brand."[35]

THE FUTURE OF SMARTPHONES

The dominance of Apple, Google, and Samsung will continue for at least a few more years. Analysts think Android will continue to be the dominant smartphone platform globally. Android's wide selection of smartphone brands, styles, and prices is difficult to beat, and the combination of Lenovo and Motorola will give Android another strong global device maker besides Samsung. iOS will keep the number-two smartphone platform spot. "It will take a fundamental change in the basis of competition to uproot either [Android or iOS] from their positions," stated VisionMobile in a 2013 report.[36] "Contenders [will have] to compete not directly but [will have to] challenge the control points of modern ecosystems: app development, distribution, and consumption of apps."[37] Windows Phone is trying to battle Android/iOS head-on, which leaves it "competing against the odds," according to VisionMobile.[38] Nevertheless, now that it is the clear number-three smartphone platform, analysts expect Windows Phone to make gains. IDC predicts that Android will account for 76 percent of the global smartphone market in 2018, iOS 14 percent, and Windows Phone 7 percent.

There may be more movement at the top in the smartphone device business than in smartphone platforms. While Samsung and Apple are expected to remain the largest smartphone makers for the near future, Motorola's carrier relationships and design talent will help Lenovo gain ground. Lenovo CEO Yang told *Fortune* that after it completes its Motorola acquisition, Lenovo intends to "compete in the full range" of smartphones, from entry-level to premium phones, and that its "mission" is to surpass Apple and Samsung.[39]

In pursuit of an edge in the ongoing smartphone wars, companies are bolstering their ecosystems. VisionMobile analyst Vakulenko says smartphone app ecosystems, having spread to tablets and Internet-connected smart TVs, are now moving to encompass car infotainment systems and wearable gadgets. Since all these devices run apps, they both enlarge and reinforce smartphone companies' app ecosystems.

A new type of device—which could be a wearable gadget or something yet to emerge—will eventually supplant smartphones, but at least some leading technology thinkers don't see that happening anytime soon. Cellphone pioneer Martin Cooper says, "We're still in the very earliest stages of smartphones." Jeff Hawkins, the father of the PDA, says, "I think we're going to be living with smartphones for a long time. They're going to get better, but I don't see them getting radically different." Dave Blakely, head of technology strategy at IDEO, the notable design and innovation consulting firm, believes smartphones will undergo "a thousand little improvements" in coming years, but he also believes that "there's no evidence that smartphones will be less central to our lives in the near future than they are now." Blakely says smartphones will endure because they are able to morph into many different devices: "Smartphones are general-purpose computing devices with personalities that can be fundamentally altered through software downloads—the whole screen can become something different."

Blakely has a unique view into the future of smartphones, since he spends his days studying emerging technologies and advising clients on technology strategy. He says smartphone users can look forward to physically customizing their phones and receiving guidance from perceptive apps. Devices called "appcessories" are one smartphone trend on Blakely's radar. These are physical devices that people attach to their smartphones to customize them, often for health purposes, and have accompanying apps that analyze and share the information they collect. For instance, an appcessory inhaler could plug into a smartphone to record a person's health data and upload it to the cloud. So could a thermometer. Appcessory users could then form a group that shares aggregate health data, yielding insights such as advance warnings of allergy outbreaks or flu epidemics.

There are some appcessories on the market today. Most are fitness gadgets that measure a user's heart rate while exercising. They typically cost close to $100 and employ Bluetooth to wirelessly communicate with their users' smartphones. Blakely says appcessory adoption would increase if manufacturers cut their specs to make them more affordable. Plug-in appcessories don't need their own batteries and processors, because they can piggyback on users' smartphones for power. The oral cancer detector OScan and the plug-in StethoMike stethoscope could be considered appcessories, though they are not meant for consumer use.

Hardware innovation isn't limited to smartphone accessories; phone design is also morphing to better suit users' needs. One of the most eagerly awaited developments is a smartphone display so flexible it can be rolled up; such devices would rival tablets and laptops in functionality, at least in terms of consuming content. In January 2013, Samsung released a concept video that shows a device approximately the length of a pen and about two to three times its width. When the user receives a notification, he presses a button that ejects a tablet display. Other Samsung concept videos depict displays that fold in half, like a book, or in thirds, like a map or a brochure. When unfolded the displays are tablets suited for viewing several types of information side by side, such as a calendar and a to-do list. When folded the displays transform into touchscreen phones.

These concepts are hampered by the limitations of today's technology. Blakely says flexible circuit boards fail at higher rates than regular ones, and flexible batteries can't store as much energy as regular ones. LG and Samsung continue to attack these problems, including through development of a flexible, carbon-based material called graphene that could be used in a range of smartphone components once it is commercialized. The two companies are expected to be the first to debut truly flexible displays, having each already introduced smartphones with slightly curved displays in 2013. Samsung has told investors it plans to bring "bended" screen devices to market in 2014 or 2015 and "foldable" screen devices in 2016.[40] Blakely thinks a rollable display is 5 to 10 years away.

Ask technologists about the future of smartphones and many will mention supersmart software assistants—like Siri on the iPhone, but smarter. Martin Cooper believes smartphones should and will offer more customized aid to users. "The intelligence in a phone is enormous; it ought to be intelligent enough to yield to your demands either in plain English or without you doing anything," Cooper says. "Phones should figure out what you need and serve you."

Frank Canova also thinks smartphones will evolve to become intelligent assistants. For example, on a day with an 80 percent chance of rain, a phone could prompt a user to wear a jacket when it senses he is leaving the house, and also remind him where the jacket is hanging. At a cocktail party, a phone could recognize when its user runs into someone he had encountered previously. "Your phone will prompt you in your ear, 'You met her a few months ago,'" says Canova. He describes this feature as "an augmentation of a person."

Blakely describes this shift as moving from apps to what he calls "agents." Apps are reactive by nature. They hold information and respond to users' requests. Siri is an app, albeit a very advanced one. In contrast, agents are intelligent advisers that proactively anticipate users' needs and offer help. They are aware of a person's history and environment, can make decisions based on this data, and may even be empowered to act on their user's behalf, based on preset rules or known preferences. An app will tell a user what time the next bus is. An agent will look at that person's calendar, the local traffic conditions, and the bus schedule, will consider the fact that the person tends to be tardy and needs nudging, and will tell him he needs to leave in five minutes in order to make a 9:00 A.M. meeting.

Technology breakthroughs are enabling the transition from apps to agents. Processors are getting more powerful, and data storage costs are declining, says Tor Björn Minde, the Ericsson research strategy chief. Machine-learning algorithms—a type of artificial intelligence used in systems that learn from data—are also advancing. As a result, companies can store more data and analyze it in more ways than ever before. Most of this work will take place in the cloud, with the phone reporting data such as a person's location and calendar updates to a remote server.

Some might say that the future is already here in the form of Google Now, a personal assistant service that is part of Google's Android and iOS search apps, or Cortana, Microsoft's recently introduced personal digital assistant for Windows Phones. Once activated, Google Now runs constantly in the phone's background and notifies users of pertinent information when it senses an upcoming event. If Google Now knows a person is about to leave on a trip, it will find the person's flight information and automatically search for delays. Before a person drives home from work, Google Now will scout for possible traffic jams and recommend alternate routes. Following its latest update, Google Now can also prompt a user when to leave in order to make a flight or another appointment. Users still need to inform Google Now of their departure location, the transportation mode they'll be using, and how early they want to arrive at their destination, but the service is approaching the cognizance level Blakely envisions.

Cortana offers similar predictive capabilities as Google Now with a dash of Siri-like personality. Cortana is powered by Bing and is the way users search the Web (by voice or by typing) in the 8.1 version of Windows Phone, which Microsoft released in 2014. Like Siri, Cortana can engage in basic conversation and carry out spoken commands, and like Google Now, it can deduce you have an upcoming trip by scanning your e-mail and will check your flight's status and the traffic to the airport, and pull up your boarding pass on your phone. Microsoft is billing Cortana as "the only digital assistant that gets to know you"[41] and has programmed it to learn users' preferences and interests to provide "tailored suggestions and proactive help."[42]

As useful as agents can be, they can also annoy users if they give bad recommendations or invade their privacy. For example, if a person likes a particular store, is near that store, and that store is currently having a sale, normally that would be a good time for the agent to notify the person of the discount. But if the person is also running late for a meeting, that fact should trump the others. An agent needs to understand that, just as a human would. Blakely calls this "the rule of and." He explains: "When an agent suggests something, every attribute needs to be correct; it needs to be the right place and time and context. If you're an intelligent agent, you need to get everything right, or you end up looking really stupid."

Privacy is the other potential pitfall for smartphone agents. Google Now asks users to share a huge amount of information, from their location to their calendars, contacts, Gmail, and other Google data.

Google Now also analyzes its users' location and Web search history, if available. Google says it won't share this information with other users or marketers without permission, but there are plenty of people who will shun the service because they consider it intrusive. "At what point do agents stop being helpful and supportive and start to become creepy?" asks Blakely. "That's something manufacturers will struggle with."

A FAIRER SMARTPHONE

The smartphone industry needs smarter policies and principles. The Dutch start-up Fairphone offers some pointers, even if it is still figuring out how best to implement them itself. As its name suggests, Fairphone makes smartphones that are aligned with fair trade's socially conscious principles, such as promoting better conditions for marginalized workers and supporting sustainable development in emerging markets. The company's mission is audacious: to amend the smartphone industry's labor, environmental, and other abuses by reforming its global supply chain. Fairphone's website tells visitors, "You can change the way products are made, starting with a single phone. Together we're . . . redefining the economy—one step at a time." [43]

Though it occupies a very small niche of the smartphone economy, Fairphone illustrates how phone makers can revamp conventional industry practices. The first Fairphone looks much like any 4.3-inch touchscreen Android handset, but it is unique because of several factors, including how it was built (with some recycled and responsibly sourced materials) and how it is being positioned (as a phone users can fully control, with a removable battery, dual SIM capability, and root access to customize the software).

Both the phone and the company were the culmination of about three years of research into the ways smartphones are produced. In 2010, a client asked Fairphone founder and CEO Bas van Abel, who is a designer by training, to create a public awareness campaign about conflict minerals. Van Abel decided to focus on smartphones, because they are the most personal of the devices that contain conflict minerals, but he soon realized that if the campaign presented no alternatives to current phones, it would be "a waste of time." Van Abel started investigating how to bring a conflict-free smartphone to the market. "Through making a phone, we could tell a story, create awareness, and find best practices," he explains from Fairphone's Amsterdam office. Van Abel also wanted to foment change by becoming part of the smartphone ecosystem rather than standing outside it. That led to the decision to establish Fairphone as a company rather than as a nonprofit.

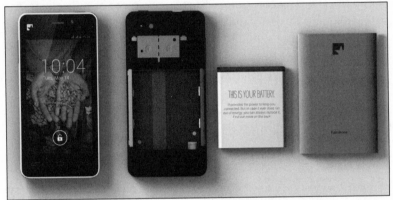

Fairphone's first smartphone, complete with dual SIM and removable battery (*Fairphone*)

Fairphone incorporated as a social enterprise in January 2013. "We have a social mission, and we're using a commercial model to get there," says van Abel. Fairphone's first phone shipped about a year later, to global interest. The start-up originally anticipated selling 5,000 phones in its first production batch. It ended up preselling 25,000 before it even produced them, and it is currently preparing a second batch of 35,000 phones.

In some ways, Fairphone functions like any smartphone maker. It sources minerals from the DRC, manufactures and assembles its smartphones in Chinese factories, and uses Android as its operating system. But since its mission informs its methods, Fairphone is more proactive about finding alternative solutions and minimizing its negative impact. The differences start with the phone itself. Its outer casing is made of postconsumer recycled polycarbonate, to reduce the amount of carbon dioxide emitted during the phone's production. It also utilizes conflict-free tantalum for its capacitors and conflict-free tin for its soldering paste. Fairphone's fairness extends to its phone's production and software, too. After selecting its manufacturing partner, Changhong, Fairphone hired TAOS Network, an independent, China-based social assessment organization, to regularly audit Changhong's factory, which is located in Chongqing, China.

Perhaps the most revolutionary aspect of Fairphone is the way it communicates with consumers. Fairphone aims to be radically transparent about its business and pricing decisions. The start-up regularly publishes blog posts about its work, from how it selected Changhong to how it is sourcing its raw materials and components. For Fairphone, transparency includes detailing how it spends its money. On its website it posted cost breakdown data that outline its expenses and specify how much of its phone's final price comes from product costs (design, engineering, manufacturing, assembly, etc.) versus operational costs (company salaries, marketing, etc.) and miscellaneous others.

Of course, 20-person start-ups are free to operate and experiment in ways massive public companies can't. Fairphone is not bound by outside investors, carriers, or retailers. It has no venture capital funding. It raised seed money from a private Dutch investor and produced its first batch of phones using money from an online crowd-funding campaign. Fairphone is distributing some phones through carriers, but it is selling the vast majority as unlocked phones through its own site. These steps enable Fairphone to maintain its independence.

Fairphone could go still further. People often ask the company why it uses the term "fair" when its materials aren't 100 percent fairly sourced. Fairphone defines fair as practices that are open and sustainable but not necessarily fair trade. Fairphone says it will work with fair trade organizations where relevant, such as when sourcing fair-trade gold, but it also collaborates with mining initiatives, trade unions, and regular industrial companies. Says van Abel: "There are all kinds of levels of fairness. . . . Fairphone is more about having a discussion about these systems that consumers are part of than nailing down something that's 100 percent fair."

Fairphone has also made mistakes and received criticism, even from its supporters. The feedback it gets shows the complexity of producing a fair smartphone, with both tech-savvy and socially conscious consumers quick to reproach Fairphone if it doesn't meet their lofty standards. Commenters have griped that Fairphone's price ($430 to $450) is too high and its specs are too low (no 4G LTE connectivity or NFC, for example), with one person protesting, "Nobody wants a super-fair, 325-Euros, 'grandma' phone."[44]

Fairphone supporters have also pushed it to do more to improve its partners' fair labor practices due to the disturbing stories they've heard about iPhone workers in China. Unfortunately, labor is one of the areas where Fairphone has yet to make dramatic progress because of inherent difficulties. TAOS Network's initial, August 2013 audit of Changhong's Chongqing factory turned up a 15-year-old worker. Fairphone says Changhong has since improved its employment age-verification procedures as well as its fire-safety measures, and started providing free lunches in its cafeteria and posting copies of local labor laws in its facilities. Other Changhong labor conditions simply match the Chinese industry standard—such as paying workers minimum wage, plus extra for overtime—or are only slightly better. When Fairphone relayed this information on its blog in December 2013, commenters said they understood factory reform would take time but also admitted they were "ashamed" and "disappointed" to learn of these conditions.[45]

As Fairphone employees readily admit, the start-up is still finding its way. "You can't make a phone completely fair within a year," says van Abel. "This is a movement, a first step in a certain direction with the right mind-set." To elevate labor conditions in Changhong's factory, Fairphone is establishing a worker welfare fund to which Fairphone will contribute $2.50 per phone sale, and Changhong will contribute another $2.50. Once the fund is set up, Changhong will hold elections so workers can choose representatives who will decide how the fund is spent. Fairphone says the money could be used for anything from "bonus payouts and leisure activities to skill-training, personal development, and team-building [workshops]."[46]

Of course, giving workers say over the disbursement of a bonus pool of money is not equivalent to letting them negotiate core issues such as pay and work hours with top management. Fairphone has promised to eventually pay its manufacturing workers a living wage, not just minimum wage, and it is hoping the fund and related elections will give workers a channel for representation and teach them collective bargaining skills. Fairphone's plan is optimistic, given Chinese labor conditions, and Fairphone is enlisting labor experts from academia and NGOs to help formulate its strategy.

Fairphone's e-waste strategy is also unusual. In addition to partnering with electronics resellers to establish a smartphone takeback and recycling system in Europe, developing repair manuals with iFixit, and selling replacement parts on its website, Fairphone is allocating $5 from every phone sale to a European foundation called Closing the Loop. The foundation used that money from the first 25,000 Fairphones to buy 70,000 beat-up cellphones in Ghana and will ship them to a Belgian smelter for recycling. Fairphone says this approach will help prevent "old, used phones from ending up in Agbogbloshie or elsewhere where they don't belong" while raising awareness in Ghana about proper recycling through local partners who collect the phones.[47] Eventually Fairphone also hopes to incorporate the metals it reclaims into future Fairphones and to build a recycling infrastructure in Ghana.

All smartphone companies can take some cues from Fairphone. As a Fairphone blog commenter wrote, the project has shown the world that tens of thousands of people "are willing to seriously consider putting social and environmental aspects on the top of their priority list when spending money on expensive products."[48] The commenter added, "Fairphone makes it easy for us to tell the world how important that is to us, maybe [even in a] much easier and more powerful [way] than by voting for [political] parties."[49]

Van Abel says Fairphone's goal is to "have an impact on the smartphone industry and to create different values—human values—within the economic system." He says he has talked to Nokia about adopting each other's sustainability practices and had general discussions with other smartphone makers. "We're not against the big players," he says. "They do things we can't and we do things they can't. We just hope we can inspire them to do some things differently."

A SMARTPHONE BILL OF RIGHTS

Like Fairphone, smartphone companies should respect core principles related to user choice, health and safety, labor rights, and the environment. Articulating how smartphone users should be treated is crucial now that smartphones are becoming ubiquitous. Here, then, is a proposed smartphone users' bill of rights.

All smartphone users should have the right to:

• Understand exactly what their phones and service plans cost.

Consumers deserve more transparency about smartphone and plan pricing. Logan Abbott of MyRatePlan always computes costs over 24 months when he compares wireless plans. Carriers should make it easier for consumers to do the same. Consumers need to know their total cost of ownership (TCO), and that requires calculating their projected voice/data service expenses over the full length of their contracts and adding in their phone purchase price and activation or upgrade fees. If carriers provided TCO estimates, either through online tools or in their stores, smartphone users would better understand their short-term and long-term costs.

• Choose what software appears on their phones.

Consumer choice should extend to smartphone software. Carriers and phone makers need to stop commandeering their customers' smartphones. There's a reason consumers call the software these companies preload onto phones "bloatware" and "crapware": it clutters up the phone's screen and occupies precious storage space while adding little value. Bloatware is a lucrative practice for smartphone companies, as it lets them charge placement fees to third-party software providers and publicize their own premium apps to a wide audience. Currently, removing bloatware requires hacking a smartphone. Smartphone users shouldn't have to do this. They should be able to uninstall bloatware through their phones' regular settings.

- Legally unlock and jailbreak their phones.

The Digital Millennium Copyright Act (DMCA) should be revised to allow consumers to circumvent the software locks that companies place on their phones. Neither jailbreaking nor rooting nor unlocking involves copyright infringement and thus should not be governed by the DMCA. (Jailbreaking and rooting are subject to the same DMCA rules as unlocking but unlike unlocking were exempted by the Librarian of Congress through the end of 2015.)[50] The law should allow third-party businesses to jailbreak and unlock phones for consumers—on an individual basis or in bulk—and to build and sell related software. In an ideal world, carriers wouldn't even lock phones. Since subscribers are already bound to their carriers through their two-year service contracts or their device payments, it isn't necessary for carriers to lock their phones.

- Know how their phone use is being tracked.

Users need more data about apps and software so they can make better choices about what to install and what to bypass. App providers should tell users not only what information they are collecting but also how that information will be used. EFF digital rights analyst Rebecca Jeschke thinks the app industry should write a universal privacy standard and make app providers adhere to it. Such a standard would curtail the number of permissions an app requested and marketing disclosures it would need to list. Jeschke suggests that app providers employ or appoint in-house ethicists, who would flag possible privacy issues and remind developers to keep data collection to a minimum.

- Keep phone information private from the government and law enforcement, barring an emergency.

Stronger laws should be passed to protect smartphone user privacy. Protection should extend to all types of smartphone data, including information that users' phones transmit to carriers and content that is located on or accessible through phones. Laws should safeguard smartphone data from law enforcement agencies in the absence of a search warrant.

- Be protected from hackers through encryption.

App providers and developers should guard their users' data by adopting HTTPS encryption. This communications protocol, called "hypertext transfer protocol secure"—adds extra security to Web communications and should protect the user data that apps collect, even if hackers intercept it. Smartphones are inherently insecure, because their communication and computing features open them up to myriad types of attacks. HTTPS is a straightforward way to make smartphones less vulnerable. The "Heartbleed" bug, a programming error that recently drew attention for threatening the security of a number of popular websites as well as phones running older versions of Android, highlighted the importance of robust encryption. Dan Auerbach, the former EFF technologist, recommends that developers implement Perfect Forward Secrecy, a stronger type of encryption that would prevent attackers from passively collecting traffic and decrypting it later if another, Heartbleed-like vulnerability is discovered. He says, "Software bugs are inevitable, but they are not an excuse to leave the door open for any attacker by not attempting to build software securely."[51]

- Feel secure that their phones are not harmful to anyone's health.

Doctors and scientists are still debating whether people should worry about smartphone RF energy affecting their health. Nevertheless, it seems prudent for the FCC to modernize their SAR testing standards to include a more diverse set of dummy models and additional phone usage scenarios. Companies should also make it easier for concerned consumers to find their smartphone's SAR value. Phone makers should reduce their use of hazardous chemicals and materials, so they can limit possible damage to people's health during manufacturing and disposal. "The notion of producer responsibility should extend to a smartphone's full lifecycle, from the chemicals that go into the product to its end-of-life treatment," says Barbara Kyle of the Electronics TakeBack Coalition.

- Learn phone makers' policies toward laborers and the environment.

Consumers should be able to figure out what company assembled their smartphones, where, and under what conditions. Smartphone makers should publicly list their suppliers, like Apple does, and their contract manufacturers. Consumers who care about phone makers' labor and environmental practices should be able to assess a company's track record, compare it to its competitors, and base their purchasing decisions on those facts. For many smartphone companies this will require disclosing more operational details than they currently do. Ideally, companies would report a common set of data in a standardized format.

- Use the same phone for several years without it becoming obsolete.

Phone makers should enable consumers to keep their phones as long as they want. As iFixit CEO Kyle Wiens says, "So much effort is put into new smartphones, and then we toss them after twelve months. I want smartphone companies to continue pushing the envelope, but I don't think we need to live in a world where we have to make a new phone for each person every year." Smartphones would have longer lifecycles if phone makers kept issuing software updates and security patches for them. Phone makers should also provide consumers with convenient, cost-efficient, third-party repair options. Smartphone users should be able to get quality repairs at independent shops or fix their phones themselves if they have the skills.

- Recycle their phones by returning them to the companies that produced or sold them.

Carriers and phone makers should be accountable for ethically disposing of the phones they sell or make. Smartphone users have an array of recycling options, but many of those recyclers may not be trustworthy. Consumers should be able to take their smartphones to their carriers or their phone makers and receive free recycling that meets the highest social and environmental standards. Smartphone companies should hire certified recyclers to ensure their used phones are processed responsibly and not dumped as e-waste or sent abroad for dismantling.

These recommendations are feasible in today's smartphone economy and applicable to any user, regardless of phone type, brand, or carrier. The list is not ranked, as different consumers have different priorities, but overall, this bill of rights indicates a need for smartphone companies to be more transparent, to better protect users, and to increase user choice.

The large smartphone companies will never be as open and honest as Fairphone. They function in an incredibly competitive, litigious industry, and most are public companies that have to impress financial analysts and satisfy investors. Even so, smartphone companies can and should be more responsive to users' needs. Smartphone users may not be asserting their full rights yet, but as smartphones pervade every facet of culture, society, and business, people will demand even more of them and the companies that make, sell, and support them.

Acknowledgments

I interviewed many people for this book and I am grateful to all of them for sharing their thoughts. My thanks to: Logan Abbott, Stan Aronow, Dan Auerbach, Russell Belk, Jeff Bell, Dave Blakely, Marcus Bleasdale, Gregory Bridgett, Abigail Sarah Brody, Frank J. Canova Jr., Debby Chan, James Clements, Ed Colligan, Rikke Constein Gerstein, Martin Cooper, Tom Cox, Geoffrey Crothall, Horace Dediu, Jason Dedrick, Chris DeSalvo, Donna Dubinsky, Todd Dunphy, Joel Engel, Hanni Fakhoury, Amina Fazlullah, Hunter Foy, Richard Frenkiel, Israel Ganot, Saskia Harmsen, Jeff Hawkins, Parker Higgins, Sheldon Hochheiser, Jason Hong, Richard Howard, Hwang Tae-hee, Rebecca Jeschke, Tomihisa Kamada, Deepa Karthikeyan, Sina Khanifar, Kim Yoo-chul, Jari Kiuru, Barbara Kyle, Henry Lai, Laurent Le Pen, Sasha Lezhnev, Li Qiang, Pui Kwan Liang, Nigel Linde, Greg Linden, Andrew Ling, Kevin Lloyd, Ollie Lo, Marina Lu, David E. Martin, Neil Mawston, Patricia Mechael, Jerry Merckel, Jim Mielke, Joe Mier, Tor Björn Minde, Florian Mueller, Paul Mugge, Ryuji Natsuumi, Michael O'Hara, Craig Palli, Chang-Min Park, Julie Pohlig, Dipankar Raychaudhuri, Karen Rudnitski, Nils Rydbeck, Jake Saunders, Joel Selanikio, Eiji Shintani, Kevin Slaten, Aaron Smith, Mark Turnage, Joel Urano, Michael Vakulenko, Bas van Abel, Tom Waldner, James Wang, Tricia Wang, Marc Weber, Kyle Wiens, Rob Williamson, Gary Wisgo, Mark Wyszomierski, Kimberly Young, and Michael Zhao.

For coordinating these interviews, contributing research or art, or otherwise helping me, I would like to thank: Chelsea Alexander, Randy Atkins, Alex Bucher, Eugene Choi, Kathleen Granchelli, Ursula Herrick, Liz Maxfield, Ami Min, Ryan Min, Sunghee Moon, Judy Purcell, Daniel Woyke, and Kathy Egan Wummer.

My thanks, also, to the team at The New Press, especially my editor, Ben Woodward; to Jon Bruner for making introductions; and to Azzurra Cox for kicking off this project.

Notes

Chapter 1: From the Simon to the BlackBerry

1. This book is primarily based on interviews. Any quotes without appended notes are from interviews. Many of the people I interviewed have also written articles, blog posts, and reports. When I am quoting a written document, I indicate that with a footnote.
2. John Schneidawind, "Big Blue Unveiling," *USA Today*, November 23, 1992.
3. Ira Sager, "Before iPhone and Android Came Simon, the First Smartphone," *Bloomberg Businessweek*, June 29, 2012.
4. "BellSouth, IBM Unveil Personal Communicator Phone," *Mobile Phone News*, November 8, 1993.
5. Dave Webb, "Intel Gets Set to Bare Its Dice," *Electronic Buyers' News*, May 30, 1994.
6. Jeffrey Schwartz, "Paging Service to Forward E-Mail," *CommunicationsWeek*, October 3, 1994.
7. Brian Nadel, "Simon: More Than a Phone, Less Than a PC," *PC Magazine*, October 25, 1994.
8. Chris O'Malley, "Unwired: The Next Generation of Communications Gear," *Popular Science*, April 1994.
9. "BellSouth Cellular Offers Voice/Data Field Service System for Simon," *Wireless Data News*, March 22, 1995.
10. Engel and Frenkiel earned the National Medal of Technology in 1994 and the National Academy of Engineering's Draper Prize in 2013 for their pioneering cellular system work.
11. Richard H. Frenkiel, *Cellular Dreams and Cordless Nightmares: Life at Bell Laboratories in Interesting Times*, 2009.
12. Youssef Ibrahim, "Finland Suddenly Takes Over as World's Most Wired Country," *New York Times*, February 1, 1997.
13. The chips were placed on opposite sides of the layered board and conduits were drawn on the inner layers.
14. Chris Ward, "Office for the Pocket Arrives," *The Times*, March 27, 1996.
15. Eric J. Savitz, "Slick Performer: Geoworks, Long a Bit Player in Operating Systems, Aims for a Starring Role," *Barron's*, May 20, 1996.
16. William Boston, "German Minister Says Buba Curbs Innovation," Reuters, March 13, 1996.

17. Nigel Powell, "For £950, a Palmtop with Phone Attached," *The Times*, August 14, 1996.

18. Susan Deemer, "Marketing & Media," *Orange County Business Journal*, October 19, 1998.

19. Hilary Barnes, "Nokia Upbeat on Sales of Handset," *Financial Times*, August 31, 1996.

20. Victor Keegan, "New Ericsson Is Smarter Than the Average Mobile," *The Guardian*, October 5, 2000.

21. William G. Phillips, ed. "1999 Best of What's New," *Popular Science*, December 1999.

22. Walter S. Mossberg, "This Digital Combo Starts with Organizer and Adds a Cell Phone," *Wall Street Journal*, November 2, 2000.

23. Edward C. Baig, "Phoning from a PDA Is at Hand," *USA Today*, November 29, 2000.

24. Henry Fountain, "A Cell Phone That Fits on the Back of an Organizer," *New York Times*, December 28, 2000.

25. Rod McQueen, *BlackBerry: The Inside Story of Research in Motion* (Toronto: Key Porter Books Limited, 2010), 28–9.

26. McQueen, *BlackBerry*, 171.

27. "Why the BlackBerry Wasn't a Strawberry," *The Times*, June 6, 2012.

28. "The New BlackBerry 5810 Wireless Handheld," Research In Motion press release, 2002.

29. Jason Brooks, "BlackBerry 5810 Heeds Call for Fewer Gadgets," *eWEEK*, May 13, 2002.

30. Bruce and Marge Brown, "RIM BlackBerry 5810: Not-So-Convenient Combo Communicator," *PC Magazine*, May 13, 2002.

31. Stephen Manes, "Strange Fruit," *Forbes*, June 10, 2002.

32. Lydia Zajc, "Small Canadian E-mail Pager Firm Takes on Big Rivals," Reuters, July 7, 1999.

33. Kevin Maney, "BlackBerry: The 'Heroin of Mobile Computing,'" *USA Today*, May 7, 2001.

34. "Handspring Unites Phone, Messaging and PDA in New Treo Communicator," Handspring press release, October 15, 2001.

35. Edward C. Baig, "Handspring Treo Does Communicate," *USA Today*, December 5, 2001.

36. Rob Pegoraro, "Even the Best of the Phone-PDA Combos Aren't Good Enough," *Washington Post*, January 27, 2002.

37. Jessica Dolcourt, "Smartphones Then and Now," July 6, 2001, CNET.

38. Steve Hamm, "Microsoft's Future," *Businessweek*, January 8, 1998.

39. Liz Vaughan-Adams, "Mobile Operators Looking to Ring Up Christmas Bonanza with New Services," *The Independent*, October 23, 2002.

40. David Pringle, "Can One Phone Do It All?" *Wall Street Journal Europe*, November 1, 2002.

41. Chana R. Schoenberger, "Not-So-Dumb Phones," *Forbes*, April 28, 2003.
42. "Microsoft Announces Windows Mobile, A Strategic Addition to the Windows Brand Family," Microsoft press release, June 23, 2003.
43. Eriko Amaha, "A La i-mode," *Far Eastern Economic Review*, August 12, 1999.
44. Tim Larimer, "What Makes DoCoMo Go," *Time*, November 27, 2000.
45. Michiyo Nakamoto, "In Japan, Cellphones Open Window to Online Services," *Financial Times*, July 5, 1999.
46. Benjamin Fulford, "DoCoMo Phone Home," *Forbes*, May 14, 2001.
47. Yuri Kageyama, "Japan DoCoMo's Ambition," Associated Press, February 28, 2002.
48. A standard for a short-range radio technology that connects smartphones to other devices, such as headsets and headphones.
49. A wireless network technology that connects smartphones to the Internet.
50. A radio system that uses satellite signals to determine a smartphone's exact location.

Chapter 2: Apple, Google, Microsoft, and Samsung

1. Ryan Block, "Live from Macworld 2007: Steve Jobs Keynote," Engadget, January 9, 2007.
2. Walter Isaacson, *Steve Jobs* (New York: Simon & Schuster, 2011), 465.
3. Ibid., 465–66.
4. Joe Mullin, "Apple v. Samsung: Apple Says Samsung Is Free-riding on $1 Billion in Marketing," Ars Technica, August 3, 2012.
5. "FingerWorks Announces a Gesture Keyboard for Apple PowerBooks." Finger-Works press release, January 27, 2004.
6. Dan Mangan, "Apple Cells It," *New York Post*, January 10, 2007.
7. Rachel Konrad, "Apple Launches iPhone, But Can It Reinvent Itself as Consumer Electronics Company?" Associated Press, January 10, 2007.
8. John Markoff, "Chiefs Defend Slow Network for the iPhone," *New York Times*, June 29, 2007.
9. Bobbie Johnson, "iCame, iSaw, iPhoned," *The Guardian*, November 5, 2007.
10. "AT&T expands WiFi," *Total Telecom*, May 22, 2006.
11. May Wong, "Apple's iPhone Stirs Up Rivals, Who Question 'Revolutionary' Claim," Associated Press, February 1, 2007.
12. "Microsoft CEO Ballmer Laughs at Apple iPhone," MacDailyNews, January 17, 2007.
13. Tae-gyu Kim, "Samsung Expects iPhones Not to Affect US Sales," *Korea Times*, January 12, 2007.
14. "Nokia Sees Apple's iPhone Stimulating Mobile Mkt," Reuters, January 25, 2007.
15. Peter Nowak, "BlackBerry Maker Brushes Off iPhone," *National Post*, June 29, 2007.
16. "Palm, Inc. Analyst and Investor Day" transcript, April 10, 2007.

17. "The iPhone Is No Smartphone, Says ABI Research," ABI press release, January 25, 2007.

18. Isaacson, *Steve Jobs*, 501.

19. John Markoff, "Phone Shows Apple's Impact on Consumer Products," *New York Times*, January 11, 2007.

20. Isaacson, *Steve Jobs*, 501.

21. Richard Martin, "It's Sexy and Cool, But Apps Are iPhone's Shortcoming," *InformationWeek*, June 18, 2007.

22. Daniel Turner, "Apple Promises Third-Party SDK for iPhone, iPod Touch," *eWEEK*, October 17, 2007.

23. "Apple Introduces the New iPhone 3G," Apple press release, June 9, 2008.

24. Ryan Block, "Steve Jobs Keynote Live from WWDC 2008," Engadget, June 9, 2008.

25. "Handango Secures Additional $9.5 Million in Venture Funding to Expand Its Growth Initiatives in the Global Smartphone Content Market," Handango press release, March 3, 2008.

26. Jefferson Graham, "App Store for iPhone Already a Hit with Developers," *USA Today*, July 20, 2008.

27. "iPhone App Store Downloads Top 10 Million in First Weekend," Apple press release, July 14, 2008.

28. Edward C. Baig, "Apple's New App Store for iPhone Stuff Is Addictive," *USA Today*, July 17, 2008.

29. John Markoff, "I, Robot: The Man Behind the Google Phone," *New York Times*, November 4, 2007.

30. Om Malik, "Danger Looms Large for the BlackBerry," *Red Herring*, June 15, 2001.

31. Jay Alabaster, "Android Founder: We Aimed to Make a Camera OS." PCWorld, April 16, 2013.

32. Ashkan Karbasfrooshan, "Top 10 Greatest U.S. Digital Media M&A Deals of All Time," TechCrunch, October 15, 2011.

33. David Smith, "The Future for Orange Could Soon Be Google in Your Pocket," *The Observer*, December 17, 2006.

34. Scott Kirsner, "Introducing the Google Phone," *Boston Globe*, September 2, 2007.

35. Jessica Guynn, "GPhone, If It Really Exists, May Revolutionize Industry," *Los Angeles Times*, September 9, 2007.

36. "Industry Leaders Announce Open Platform for Mobile Devices," Open Handset Alliance press release, November 5, 2007.

37. Ibid.

38. Ibid.

39. Simon Dumenco, "Pop Pick," *Advertising Age*, November 12, 2007.

40. Joe Wilcox, "God Phone Meets the Devil," *eWEEK* Apple Watch blog, September 23, 2008.

41. David Pogue, "A Look at Google's First Phone," *New York Times*, October 16, 2008.
42. Elizabeth Lazarowitz, "Phone Fight!" *New York Daily News*, September 24, 2008.
43. Jack Schofield, "Google Android—So Far, a Haven for Useless Apps," *The Guardian* technology blog, October 23, 2008.
44. Fred Vogelstein, "The Day Google Had to 'Start Over' on Android," *The Atlantic*, December 18, 2013.
45. Bonnie Cha and Nicole Lee, "T-Mobile G1 Review," CNET, October 21, 2008.
46. Federal Communications Commission, "The Global Internet at a Crossroads," Council on Foreign Relations, Washington, D.C., November 20, 2012.
47. Kate Bevan, "Microsoft's Outsmarted Smartphone Rings in the New," *The Guardian*, August 6, 2009.
48. Dan Frommer, "What If Microsoft Had Bought BlackBerry in 2009 Instead?" SplatF, September 3, 2013.
49. "BlackBerry Board of Directors Announces Exploration of Strategic Alternatives," BlackBerry press release, August 12, 2013.
50. Steven M. Davidoff, "What a Big Investment Says About BlackBerry's Endgame," *New York Times* DealBook blog, November 4, 2013.
51. Tony Michell, *Samsung Electronics and the Struggle for Leadership of the Electronics Industry* (Singapore: John Wiley & Sons [Asia] Pte. Ltd., 2010), 5.
52. Ibid., 155.
53. Brian X. Chen, "Samsung Challenges Apple's Cool Factor," *New York Times*, February 10, 2013.

Chapter 3: The Smartphone Wars

1. Brian X. Chen and Nick Wingfield, "A Challenge in iPhone's Backyard," *New York Times*, March 16, 2013.
2. Ina Fried, "Samsung Galaxy S IV Debuts at Radio City Music Hall," All Things D, March 14, 2013.
3. Tiernan Ray, "Not Quite Ready for Broadway," *Barron's*, March 16, 2013.
4. Isaacson, *Steve Jobs*, 512.
5. Liz Gannes, "The Six Juiciest Documents from the Apple-Samsung Trial This Week," Recode, April 5, 2014.
6. "Apple Inc. Fiscal First Quarter 2009 Earnings Call Transcript," Seeking Alpha, January 21, 2009.
7. Matt Hartley, "Google's Andy Rubin: The Man Who Redrew the Mobility Map," *Financial Post*, August 7, 2011.
8. Steven Levy, "The Inside Story of the Moto X," *Wired*, August 1, 2013.
9. "Google's CEO Discusses Q3 2012 Results—Earnings Call Transcript," Seeking Alpha, October 18, 2012.
10. Ina Fried, "Google's Andy Rubin Gives a Flash of Tablet Future," All Things D, December 6, 2010.

11. "Google's CEO Discusses Q3 2012 Results—Earnings Call Transcript."

12. Greg Sterling, "What to Make of Google's $8 Billion 'Mobile Run Rate' Figure," Marketing Land, October 19, 2012.

13. Brian Womack, "Google's YouTube Triples Mobile Sales Amid Wireless Shift," Bloomberg, June 5, 2013.

14. Benjamin Travis, "Android vs. iOS: User Differences Every Developer Should Know," comScore blog, March 13, 2013.

15. Tiernan Ray, "Sprint: Street Mulls PCS Rumors, iPhone Impact," *Barron's*, February 27, 2012.

16. Sam Grobart, "Tim Cook: The Complete Interview," *Bloomberg Businessweek*, September 20, 2013.

17. Satya Nadella, "A Cloud for Everyone, on Every Device," Official Microsoft blog, March 27, 2014.

18. "Accelerating Growth" PowerPoint presentation, Microsoft website, September 3, 2013, http://www.microsoft.com/en-us/news/download/press/2013/strategic rationale.pdf.

19. Philip Elmer-DeWitt, "Transcript: Apple CEO Tim Cook at Goldman Sachs," *Fortune*.

20. Michell, *Samsung Electronics*, 151.

21. Spencer Ante, "Apple Closes U.S. Ad-Spending Gap with Samsung," *Wall Street Journal* Digits blog, April 8, 2014.

22. Suzanne Vranica, "Behind the Preplanned Oscar Selfie: Samsung's Ad Strategy," *Wall Street Journal*, March 3, 2014.

23. Miyoung Kim, "Samsung's Marketing Splurge Doesn't Always Bring Bang-for-Buck," Reuters, November 27, 2013.

24. Brian X. Chen, Nick Wingfield, James Kanter, and Kevin J. O'Brien, "Europe Weighs iPhone Sale Deals with Carriers for Antitrust Abuse," *New York Times*, March 21, 2013.

25. Horace Dediu, "Who's Next?" Asymco blog, September 4, 2013.

26. Aapo Markkanen, "Galaxy S4, a Launch Pad for Samsung's Great OS Escape," ABI Research blog, March 15, 2013.

27. Miyoung Kim and Se Young Lee, "Samsung Executive Says Galaxy S5 to Outsell S4, Sees Second Quarter Rollout for Tizen Phone," Reuters, April 16, 2014.

28. Jessica E. Lessin, Lorraine Luk, and Juro Osawa, "Apple Finds It Difficult to Divorce Samsung," *Wall Street Journal*, July 1, 2013.

29. Ibid.

30. Jon Fingas, "Apple Rumored to Need Samsung for Some A8 Chip Production," Engadget, September 29, 2013; "Samsung, TSMC to Share Apple 14/16nm Chip Orders," *Digitimes*, December 18, 2013.

31. Apple v. Samsung complaint, April 15, 2011, http://images.apple.com/pr/pdf /110415samsungcomplaint.pdf.

32. Apple v. Samsung, United States District Court, N.D. California, San Jose Di-

vision, July 1, 2012, http://www.leagle.com/decision/In%20FDCO%202012070
2714.

33. Ashby Jones and Jessica E. Vascellaro, "Apple v. Samsung: The Patent Trial of the
Century," *Wall Street Journal*, July 24, 2012.

34. Florian Mueller, "In 49 Months of Holy War, Apple Has Not Proved that It Owns
Any Feature Other Than Rubber-Banding," FOSS Patents blog, April 2, 2014.

35. Florian Mueller, "Retrial Jury Awards Apple $290 Million, Total Damages
in Case Against Samsung: $929 million," FOSS Patents blog, November 21,
2013.

36. Joel Rosenblatt, "Apple Seeks Android Source Code Records in Samsung Suit,"
Bloomberg, May 8, 2013.

37. Isaacson, *Steve Jobs*, 512.

38. Roger Cheng and Greg Sandoval, "How Google's Stealth Support Is Buoying
Samsung in Apple Fight," CNET, August 10, 2012.

39. Gavin Clarke, "How Did Microsoft Get to Be a $1.2bn Phone Player? Hint: NOT
Windows Phone," The Register, July 31, 2013.

40. Liam Tung, "Microsoft Is Making $2bn a Year on Android Licensing—Five
Times More Than Windows Phone," ZDNet, November 7, 2013.

41. David Drummond, "When Patents Attack Android," Official Google blog, August 3, 2011.

42. Joe Mullin, "Patent War Goes Nuclear: Microsoft, Apple-Owned 'Rockstar' Sues
Google," Ars Technica, October 31, 2013.

43. Numbers current as of August 2013.

44. Bradley Johnson, "Big U.S. Advertisers Boost 2012 Spending by Slim 2.8% with
a Lift From Tech," *Advertising Age*, June 23, 2013.

45. Mike Shields, "AT&T Sues Verizon Wireless," *Adweek*, November 4, 2009.

46. Rita Chang, "Verizon vs. AT&T: Blistering Battle Raging Over Map," *Advertising
Age*, November 30, 2009.

47. Chris Welch, "T-Mobile CEO Pokes the Bears, Calls AT&T's Network 'crap' and
Mocks Verizon," The Verge, January 9, 2013.

48. Jay Yarow, "Google Is Reportedly Trying to Get a Bigger Slice of Android App
Revenue," Business Insider, June 28, 2013.

49. "Telefónica to Promote and Foster Sales of Windows Phone 8 in the United
Kingdom, Germany, Spain, Mexico, Brazil and Chile," Telefónica press release,
June 26, 2013.

50. MG Siegler, "The Google-Free iPhone," TechCrunch, August 6, 2012.

51. Dominic Basulto, "Remember What Algorithms Were to Search? That's What
Maps Are to Mobile," *Washington Post* Innovations blog, September 26, 2012.

52. David Howard, "The Limits of Google's Openness," Official Microsoft blog, August 15, 2013.

53. Tom Warren, "Inside the Bitter YouTube Battle Between Microsoft and Google,"
The Verge, August 16, 2013.

54. Lance Whitney, "Microsoft Neuters YouTube Windows Phone App," CNET, October 8, 2013.

55. Jenna Wortham and Nick Wingfield, "Microsoft Is Writing Checks to Fill Out Its App Store," *New York Times*, April 5, 2012.

56. Blair Hanley Frank, "Surprise, You've Got a Windows Phone App!" GeekWire, February 20, 2014.

57. "App Annie Index: 2013 Retrospective" report, App Annie, January 30, 2014.

58. "Developer Economics Q1 2014" report, VisionMobile, February 2014.

59. "Developer Economics 2012" report, VisionMobile, June 2012.

60. Panos Papadopoulos, "Rise of the Mega SDK Vendors in Mobile," VisionMobile blog, July 2, 2013.

61. Carter Thomas, "Using Bots for App Store Rankings," Blue Cloud Solutions blog, February 7, 2014.

62. Christina Warren, "28 Days of Fame: The Strange, True Story of 'Flappy Bird,' " Mashable, February 10, 2014.

63. Nguyen told *Rolling Stone* that media scrutiny and criticisms about his game's addictiveness were the main reasons he pulled the game. (David Kushner, "Flight of the Birdman," *Rolling Stone*, March 11, 2014.)

64. Not all cross-promotion networks let developers pay for downloads, but Vision-Mobile doesn't specifically track "appayola" usage. (Developer Economics Q1 2014 report.)

65. Sarah Perez, "Apple's App Store Rankings Algorithm Changed to Consider Ratings, and Possibly Engagement," TechCrunch, August 23, 2013.

66. "Transnational Organized Crime in East Asia and the Pacific: A Threat Assessment" report, United Nations Office on Drugs and Crime, April 2013.

67. "New Study Shows Counterfeit and Substandard Phones Significantly Reduce Consumer Experience," Mobile Manufacturers Forum press release, October 21, 2011.

68. David Meyer, "Samsung: We don't region-lock our phones (apart from when you first activate them)," Gigaom, September 27, 2013.

Chapter 4: Assembling a Smartphone

1. Tim Bradshaw and Sarah Mishkin, "HTC Hints at Life Beyond Smartphones," *Financial Times*, October 20, 2013.

2. Adam Satariano, "Apple Trial Offers Glimpse of Kitchen-Table Product Design," Bloomberg, July 31, 2012.

3. "From Paper to Stone—How Samsung Product Designs Are Inspired," Samsung Village blog, September 20, 2011.

4. Farhad Manjoo, "Dynamic Duos: Samsung on Global Design Influences," *Fast Company*, October, 2013.

5. Ibid.

6. Ibid.

7. Kevin Bostic, "Samsung Design Chief Talks Plastic and Software, Says Future Is in Devices with 'Souls,' " Apple Insider, March 16, 2013.

8. Brian X. Chen, "Samsung Emerges as a Potent Rival to Apple's Cool," *New York Times*, February 11, 2013.

9. Andy Reinhardt, "Steve Jobs: 'There's Sanity Returning,' " *Businessweek*, May 25, 1998.

10. Nicole Perlroth and Nick Wingfield, "Design and Drama Mark First Day in Apple-Samsung Trial," *New York Times* Bits blog, July 31, 2012.

11. Hayley Tsukayama, "Galaxy Note: Big Screen Browsing, But Is It Practical?" *Washington Post*, February 22, 2012.

12. "It May Be a 'Brick,' But the Galaxy Note Is Still a Smartphone," *The Mercury* (South Africa), February 8, 2012.

13. "Is It a Bird? Is It a Plane? No, It's the New Samsung Galaxy 'Phablet'!" *Irish Independent*, March 13, 2012.

14. Ina Fried, "Apple Literally Designs Its Products Around a Kitchen Table," All Things D, July 31, 2012.

15. Isaacson, *Steve Jobs*, 519.

16. Charles Arthur, "Marko Ahtisaari: Smartphone Evolution Is Only Just Beginning," *The Guardian*, January 31, 2012.

17. "Protection of Semiconductor Chip Products," Copyright Law of the United States of America.

18. Kenneth Kraemer, Greg Linden, and Jason Dedrick, "Capturing Value in Global Networks: Apple's iPad and iPhone," July 2011.

19. John Gruber, "The New Apple Advantage," Daring Fireball blog, September 9, 2011.

20. Daniel Eran Dilger, "Apple to Spend $10 Billion on Innovation, Expansion in 2013," Apple Insider, January 24, 2013.

21. Horace Dediu, "The Bank of Apple: Using Capital to Ensure Additional Capacity and Supply," Asymco blog, January 23, 2011.

22. Daniel Eran Dilger, "Apple Signs $578M Sapphire Deal with GT Advanced Technology," AppleInsider, November 4, 2013.

23. Debra Hofman, Stan Aronow, and Kimberly Nilles, "The Gartner Supply Chain Top 25 for 2013" report, Gartner, May 22, 2013.

24. Miyoung Kim, "A Stretched Samsung Chases Rival Apple's Suppliers," Reuters, May 17, 2013.

25. Bradshaw and Mishkin, "HTC Hints at Life Beyond Smartphones."

26. "Supplier Responsibility 2014 Progress Report," Apple.

27. "Frequently Asked Questions, Conflict Minerals," Dodd-Frank Wall Street Reform and Consumer Protection Act, May 30, 2013.

28. Zoe Lewis and Sasha Lezhnev, "National Geographic on Conflict Minerals: Opportunity to Grow the Clean Minerals Trade in Congo," Enough Project blog, September 20, 2013.

29. Simon Duke, "(Not) Made in China," *Sunday Times*, December 30, 2012.

30. John Chen, "BlackBerry: The Way Forward," CNBC.com, December 30, 2013.
31. Joshua Topolsky, "What We Learned from the 'Nightline' Report on Foxconn Factories," The Verge, February 22, 2012.
32. Aditya Chakrabortty, "The Woman Who Nearly Died Making Your iPad," *The Guardian*'s Comment Is Free site, August 5, 2013.
33. Pun Ngai, Shen Yuan, Guo Yuhua, Lu Huilin, Jenny Chan, and Mark Selden, "Worker-Intellectual Unity: Trans-Border Sociological Intervention in Foxconn," *Current Sociology*, January 7, 2014.
34. Frederik Balfour and Tim Culpan, "The Man Who Makes Your iPhone," *Bloomberg Businessweek*, September 9, 2010.
35. "Apple Says 'Saddened' by String of Suicides at Foxconn," Reuters, May 26, 2010.
36. John Paczkowski, "Apple CEO Steve Jobs Live at D8," All Things D, June 1, 2010.
37. "Supplier Responsibility 2011 Progress Report," Apple.
38. "Apple's Supplier Pegatron Group Violates Workers' Rights" report, China Labor Watch, July 29, 2013.
39. Ibid.
40. "Biel Crystal Investigative Report," SACOM, November 25, 2013.
41. "Samsung's Supplier Factory Exploiting Child Labor" report, China Labor Watch, August 6, 2012.
42. "An Investigation of Eight Samsung Factories in China: Is Samsung Infringing Upon Apple's Patent to Bully Workers?" report, China Labor Watch, September 4, 2012.
43. "Regarding Samsung's Action Plans to Review Working Conditions at Chinese Factories," Samsung Village blog, September 4, 2012.
44. "More Labor Abuse in Samsung Phone Factory" report, China Labor Watch, December 5, 2013.
45. "Supplier Responsibility 2014 Progress Report," Apple.
46. "Independent External Monitoring Reports," Fair Labor Association website.
47. Bill Weir, "A Trip to the iFactory: 'Nightline' Gets an Unprecedented Glimpse Inside Apple's Chinese Core," *ABC Nightline*, February 20, 2012.
48. "Final Foxconn Verification Status Report," Fair Labor Association, December 12, 2013.
49. "Supplier Responsibility 2014 Progress Report," Apple.
50. "China's Official Trade Union Still Fails to Get the Message," China Labour Bulletin, December 2, 2013.
51. "English Translation Excerpts of the Report on Foxconn Trade Union Research," SACOM website, May 1, 2013.
52. Kathrin Hille and Rahul Jacob, "Foxconn Plans Chinese Union Vote," *Financial Times*, February 3, 2013.
53. Wang Pan and Lu Qiuping, "China Exclusive: Foxconn Union Election Described as Revamp," Xinhua News, February 5, 2013.

Chapter 5: Waste: Money and Trash

1. Brian X. Chen, "Ratemizer App Offers Instant Phone-Bill Analysis," *New York Times* Bits blog, November 20, 2012.
2. "Consumer Reports: Cell Phone Plan Savings," ABC News website, November 21, 2013.
3. Thomas Gryta, "AT&T's Plan Revamp Signals the End of Voice Minutes," *Wall Street Journal* Digits blog, October 25, 2013.
4. Chetan Sharma, "U.S. Mobile Market Update—Q4 2013 and 2013," Chetan Sharma website.
5. Tammy Parker, "Sprint Slams on the Brakes for Top 5% of Data Users in Congested Areas," *FierceWirelessTech*, May 8, 2014.
6. Urvaksh Karkaria, "AT&T: No Unlimited Data Plan for You!" *Atlanta Business Chronicle*, October 16, 2010.
7. "AT&T Inc. at UBS Global Media and Communications Conference" transcript, December 5, 2012.
8. Thomas Gryta, "Inside the Phone-Plan Pricing Puzzle," *Wall Street Journal*, July 31, 2013.
9. "Ericsson Mobility Report," Ericsson website, November 2013.
10. Kevin Fitchard, "A Bird's Eye View of the AT&T-Leap Wireless Merger," Gigaom, July 15, 2013.
11. "Sprint CEO Discusses Q4 2013 Results—Earnings Call Transcript," Seeking Alpha, February 11, 2014.
12. "Orange 'Forces Google' to Pay for Mobile Traffic," Agence France-Presse, January 16, 2013.
13. "2013 Most Reputable U.S. Companies" report, Reputation Institute, http://www.rankingthebrands.com/PDF/Global%20Reputation%20Pulse%20-%20U.S.%20Top%20100%202013.pdf.
14. Brian X. Chen, "AT&T and Sprint Chiefs Display a Difference in Moods," *New York Times* Bits blog, May 9, 2012.
15. Marguerite Reardon, "AT&T Chief: We Can't Keep Doing Big Subsidies on Phones," CNET, December 10, 2013.
16. Jason Gilbert, "Carly Foulkes, 'T-Mobile Girl,' Looks Like She's Been Fired," *Huffington Post*, March 27, 2013.
17. David Pogue, "Breaking Free of the Cellphone Carrier Conspiracy," *New York Times*, April 3, 2013.
18. Roger Cheng, "T-Mobile Faces War of Words After Killing Subsidies, Contracts," CNET, March 26, 2013.
19. "Attorney General says 'No Dice' to T-Mobile's deceptive 'No-Contract' advertising," Washington State Office of the Attorney General website, April 25, 2013.
20. Ibid.
21. Amir Rozwadowski, "T-Mobile U.S. Inc.: And the Ball Keeps Rolling . . . ," Barclays Equity Research note to investors, January 9, 2014.

22. Phil Goldstein, "We're Not Having a Wireless Price War Now—But What Would One Look Like?" *FierceWireless*, March 11, 2014.

23. Brian X. Chen, "Carriers Step Up Battle for Wireless Customers," *New York Times*, January 8, 2014.

24. Amir Rozwadowski, "Smartphone Data Points: Indicative of a Maturing Upgrade Cycle," Barclays Equity Research note to investors, July 8, 2013.

25. "Exemption to Prohibition on Circumvention of Copyright Protection Systems for Access Control Technologies," *Federal Register*, Vol. 77, No. 208, Library of Congress Copyright Office, October 26, 2012.

26. R. David Edelman, "It's Time to Legalize Cell Phone Unlocking," Official White House response, March 4, 2013.

27. FCC letter to CTIA, November 14, 2013, http://transition.fcc.gov/Daily_Releases /Daily_Business/2013/db1114/DOC-324166A1.pdf.

28. Mike Masnick, "Latest Congressional Attempt to 'Fix' Mobile Phone Unlocking Just Punts the Issue Until Later," Techdirt, March 12, 2013.

29. "Criminal Offenses and Penalties," Digital Millennium Copyright Act, 17 U.S. Code § 1204.

30. Catherine Rampell, "Planned Obsolescence, as Myth or Reality," *New York Times* Economix blog, October 31, 2013.

31. John Gruber, "2013: The Year in Apple and Technology at Large," Daring Fireball blog, December 27, 2013.

32. Ibid.

33. Huabo Duan, T. Reed Miller, Jeremy Gregory, and Randolph Kirchain, "Quantitative Characterization of Domestic and Transboundary Flows of Used Electronics Analysis of Generation, Collection, and Export in the United States" report, December 2013.

34. Ivan Watson, "China: The Electronic Wastebasket of the World," CNN.com, May 30, 2013.

35. Michael Zhao, "E:Waste: Afterlife," 2011.

36. Feng Wang, Ruediger Kuehr, Daniel Ahlquist, and Jinhui Li, "E-Waste in China: A Country Report," April 5, 2013.

37. "Where Are WEEE in Africa?: Findings from the Basel Convention E-Waste Africa Programme," Secretariat of the Basel Convention, December 2011.

38. Ibid.

39. "Basel Convention on the Control of Transboundary Movements of Hazardous Wastes and Their Disposal," United Nations Environment Programme.

40. "Legislation Threatens Responsible Recycling and American Jobs," ISRI press release, July 24, 2013.

41. Sprint Buyback Program website.

42. Stela Bokun, "Buy-Back and Trade-In Programs Reduce the Weight of Subsidies for Mobile Operators" report, Pyramid Research, November 26, 2013.

43. "The Responsible Recycling ('R2') Standard for Electronics Recyclers," R2 Solutions, July 1, 2013.

44. "Activists Cite IG Study in Push for Federal Adoption of Strict E-Waste Plan," *Defense Environment Alert*, November 9, 2010.
45. Peter Boaz, "U.S.: Inmates Exposed to Toxic E-Waste," Inter Press Service, October 29, 2010.
46. "BAN Applauds LG Electronics for Responsible e-Waste Recycling," BAN blog, September 9, 2013.

Chapter 6: Health

1. "IARC Report to the Union for International Cancer Control (UICC) on the Interphone Study," Interphone website, October 3, 2011.
2. Ibid.
3. Adam Nichols, "Shock Finding: Cellphone-Cancer Link," *New York Post*, October 25, 2009.
4. "Watch That Thing by Your Head," *Globe and Mail*, May 21, 2010.
5. "Better Safe Than Sorry," *New York Times*, June 3, 2011.
6. "The Cellphone Panic," *Wall Street Journal*, June 4, 2011.
7. Devra Davis, *Disconnected: The Truth About Cell Phone Radiation, What the Industry Is Doing to Hide It, and How to Protect Your Family* (New York: Dutton, 2010), 183.
8. "Herrera Blasts Wireless Phone Industry's Attempt to Take Away San Francisco Consumers' Right to Know," Office of the City Attorney press release, October 4, 2011.
9. Davis, *Disconnected*, 82.
10. "Exposure and Testing Requirements for Mobile Phones Should Be Reassessed," United States Government Accountability Office report, GAO website, July 2012.
11. Scott Stinson, "The Cellphone Cancer Scare That Won't Go Away," *National Post*, May 21, 2010.
12. Sherry Turkle, *Alone Together: Why We Expect More from Technology and Less from Each Other* (New York: Basic Books, 2011), 161.
13. Tom Cheney, *New Yorker* cartoon, May 27, 2013.
14. Mark D. Griffiths, "Adolescent Mobile Phone Addiction: A Cause for Concern?" *Education and Health*, Vol. 31, No. 3, 2013.
15. "Frequently Asked Questions," Net Addiction website.
16. Kimberly S. Young, "Internet Addiction: A New Clinical Phenomenon and Its Consequences," *American Behavioral Scientist*, Vol. 48, No. 4, December 2004.
17. Internet Gaming Disorder fact sheet, DSM-5 website.
18. John D. Sutter, "Is 'Gaming Addiction' a Real Disorder?" CNN.com, August 6, 2012.
19. Kimberly S. Young, "CBT-IA: The First Treatment Model for Internet Addiction," *Journal of Cognitive Psychotherapy*, Vol. 25, No. 4, 2011.
20. Internet Addiction program website, Bradford Regional Medical Center website.
21. "14 Pct of Students Addicted to Smartphones: Poll," Yonhap News, July 3, 2013.

22. Jung Ha-won, "Ultra-Wired South Korea Battles Smartphone Addiction," Agence France-Presse, June 26, 2013.

23. Min Kwon, Joon-Yeop Lee, Wang-Youn Won, Jae-Woo Park, Jung-Ah Min, Changtae Hahn, Xinyu Gu, Ji-Hye Choi, and Dai-Jin Kim, "Development and Validation of a Smartphone Addiction Scale (SAS)," http://www.ncbi.nlm.nih .gov/pubmed/23468893.

24. Baek Il-hyun and Park Eun-jee, " 'Digital Dementia' Is on the Rise," *Korea Joong-Ang Daily*, June 24, 2013.

25. Kwon et al., "Development and Validation of a Smartphone Addiction Scale (SAS)."

26. Ibid.

27. Kris Newby, "New Smartphone Scans from Stanford Could Prevent Needless Oral Cancer Deaths," Stanford School of Medicine press release, April 17, 2012.

28. Luis J. Haddock, David Y. Kim, and Shizuo Mukai, "Simple, Inexpensive Technique for High-Quality Smartphone Fundus Photography in Human and Animal Eyes," *Journal of Ophthalmology*, 2013.

29. Joel Selanikio, "Four Phases of Field Data: Tech Transforming Data Collection for Development," DataDyne blog, March 21, 2013.

30. Scott Hillis, "Technology Aid for Relief Groups," Reuters, July 5, 2001.

31. The project is called PartME for Participatory Monitoring and Evaluation and is funded by two Netherlands-based organizations: the International Institute for Communication and Development (IICD), a nonprofit that promotes the use of technology as a development tool, and the Catholic Organisation for Relief and Development Aid (CORDAID).

32. "Oxfam Annual Report 2012–2013," Oxfam website.

Chapter 7: Privacy

1. "Did Your Smartphone Flashlight Rat You Out?" Carnegie Mellon University press release, January 15, 2013.

2. Flurry Personas, Flurry website.

3. Greg Sterling, "Google Replacing 'Android ID' with 'Advertising ID' Similar to Apple's IDFA," Marketing Land, October 31, 2013.

4. Adrienne Porter Felt, Elizabeth Ha, Serge Egelman, Ariel Haney, Erika Chin, and David Wagner, "Android Permissions: User Attention, Comprehension, and Behavior" report, 2012.

5. Ibid.

6. Ibid.

7. "Path Social Networking App Settles FTC Charges it Deceived Consumers and Improperly Collected Personal Information from Users' Mobile Address Books," Federal Trade Commission press release, February 1, 2013.

8. "Complying with COPPA: Frequently Asked Questions," Federal Trade Commission website, July 2013.

9. "Android Flashlight App Developer Settles FTC Charges It Deceived Consumers," Federal Trade Commission press release, December 5, 2013.

10. "Mobile Privacy Disclosures: Building Trust Through Transparency" report, Federal Trade Commission website.

11. Ibid.

12. http://www.ftc.gov/system/files/documents/public_comments/2014/02/00008 -88567.pdf.

13. Julianne Pepitone, "What Your Wireless Carrier Knows about You," CNN.com, December 16, 2013.

14. Privacy Policy, Sprint website.

15. Jessica Leber, "How Wireless Carriers Are Monetizing Your Movements," *MIT Technology Review*, April 12, 2013.

16. Glenn Greenwald, "NSA Collecting Phone Records of Millions of Verizon Customers Daily," *The Guardian*, June 6, 2013.

17. Ibid.

18. "Congressional Hearing on the National Security Agency" transcript, June 18, 2013, http://transcripts.cnn.com/TRANSCRIPTS/1306/18/cnr.04.html.

19. Siobhan Gorman, "NSA Collects 20% or Less of U.S. Call Data," *Wall Street Journal*, February 7, 2014.

20. Ellen Nakashima, "NSA Is Collecting Less than 30 Percent of U.S. Call Data, Officials Say," *Washington Post*, February 7, 2014.

21. "Liberty and Security in a Changing World: Report and Recommendations of the President's Review Group on Intelligence and Communications Technologies," December 12, 2013.

22. Ibid.

23. "Report on the Telephone Records Program Conducted under Section 215 of the USA PATRIOT Act and on the Operations of the Foreign Intelligence Surveillance Court," Privacy and Civil Liberties Oversight Board, January 23, 2014.

24. "Remarks by the President on Review of Signals Intelligence," January 17, 2014.

25. Ibid.

26. Julian Sanchez, "The Best NSA Fix Comes from the Patriot Act's Author," The Daily Beast, March 25, 2014.

27. Fourth Amendment, http://www.law.cornell.edu/constitution/fourth_amend ment.

28. "Congressional Hearing on the National Security Agency" transcript.

29. Richard Leon, Memorandum Opinion, December 16, 2013, https://ecf.dcd.us courts.gov/cgi-bin/show_public_doc?2013cv0851-48.

30. Joe Mullin, "Sounding the Alarm: Ars Speaks with Vocal NSA Critic Sen. Ron Wyden," Ars Technica, July 31, 2013.

31. Jonathan Mayer and Patrick Mutchler, "MetaPhone: The Sensitivity of Telephone Metadata," March 12, 2014, http://webpolicy.org/2014/03/12/metaphone -the-sensitivity-of-telephone-metadata/.

32. "For Second Year in a Row, Markey Investigation Reveals More Than One Million Requests by Law Enforcement for Americans Mobile Phone Data," Ed Markey press release, December 2013.

33. "Verizon Transparency Report," Verizon website, January 22, 2014.

34. "Transparency Report," AT&T website, February 18, 2014.

35. Barton Gellman and Ashkan Soltani, "NSA Tracking Cellphone Locations Worldwide, Snowden Documents Show," *Washington Post*, December 4, 2013.

36. Ashkan Soltani and Barton Gellman, "New Documents Show How the NSA Infers Relationships Based on Mobile Location Data," *Washington Post* The Switch blog, December 10, 2013.

37. Gellman and Soltani, "NSA Tracking Cellphone Locations Worldwide, Snowden Documents Show."

38. Charlie Savage, "In Test Project, N.S.A. Tracked Cellphone Locations," *New York Times*, October 2, 2013.

39. Paul Lewis, "NSA Chief Admits Agency Tracked U.S. Cellphone Locations in Secret Tests," *The Guardian*, October 2, 2013.

40. Yves-Alexandre de Montjoye, César A. Hidalgo, Michel Verleysen, and Vincent D. Blondel, "Unique in the Crowd: The Privacy Bounds of Human Mobility," Scientific Reports, March 25, 2013.

41. Bradford Pearson, "License Plate Readers Coming to Dallas," *D Magazine*, February 25, 2013.

42. Allie Bohm, "Utah Enacts Significant Location and Communications Privacy Bill," ACLU blog, April 3, 2014.

43. Chris Soghoian, "New Document Sheds Light on Government's Ability to Search iPhones," ACLU blog, February 26, 2013.

44. Catherine Crump, "Capability Is Driving Policy, Not Just at the NSA But Also in Police Departments," ACLU blog, November 1, 2013.

45. Linda Lye, "DOJ Emails Show Feds Were Less Than 'Explicit' with Judges on Cell Phone Tracking Tool," ACLU blog, March 27, 2013.

46. Jon Campbell, "LAPD Spied on 21 Using StingRay Anti-Terrorism Tool," *LA Weekly*, January 24, 2013.

47. Nathan Freed Wessler, "Local Police in Florida Acting Like They're the CIA (But They're Not)," ACLU blog, March 25, 2014.

48. John Kelly, "Cellphone Data Spying: It's Not Just the NSA," *USA Today*, December 8, 2013.

49. Ibid.

50. Hanni Fakhoury and Trevor Timm, "Stingrays: The Biggest Technological Threat to Cell Phone Privacy You Don't Know About," Electronic Frontier Foundation blog, October 22, 2012.

51. Erica Fink, "This Drone Can Steal What's on Your Phone," CNN.com, March 20, 2014.

52. Tom A. Peter, "Drones on the U.S. Border: Are They Worth the Price?" *Christian Science Monitor*, February 5, 2014.

53. "Unmanned Aircraft System MQ-9 Predator B" fact sheet, U.S. Customs and Border Protection, May 1, 2013, http://www.cbp.gov/sites/default/files/documents/predator_b_7.pdf.

54. Jennifer Lynch, "Customs & Border Protection Loaned Predator Drones to Other Agencies 700 Times in Three Years According to 'Newly Discovered' Records," Electronic Frontier Foundation blog, January 14, 2014.

55. "Consumers Will Share Personal Data . . . at a Price," Amdocs survey, June 20, 2013, http://www.amdocs.com/News/Pages/amdocs-personal-data-consumer-survey.aspx.

56. Jennifer M. Urban, Chris Jay Hoofnagle, and Su Li, "Mobile Phones and Privacy" survey, University of California, Berkeley, July 11, 2012.

57. Ibid.

58. "Survey of Young Americans' Attitude Toward Politics and Public Service: 24th Edition," Institute of Politics, John F. Kennedy School of Government, Harvard University,December4,2013,http://www.iop.harvard.edu/survey-young-americans%E2%80%99-attitude-toward-politics-and-public-service-24th-edition.

59. *Smith v. Maryland* opinion, United States Supreme Court, June 20, 1979, http://caselaw.lp.findlaw.com/scripts/getcase.pl?court=US&invol=735&vol=442.

Chapter 8: Looking Toward the Future

1. David Honig, "More Wireless Broadband Is What Consumers Want, U.S. Needs to Close the Digital Divide," *Huffington Post*, January 3, 2012.

2. Steve Largent, "CTIA-The Wireless Association Statement on President Obama's National Wireless Innovation and Infrastructure Initiative," CTIA blog, February 10, 2011.

3. Kathryn Zickuhr and Aaron Smith, "Home Broadband 2013" report, Pew Research Center's Internet & American Life Project, August 26, 2013.

4. Ibid.

5. Philip M. Napoli and Jonathan A. Obar, "Mobile Leapfrogging and Digital Divide Policy," New America Foundation, April 2013.

6. Susan P. Crawford, "The New Digital Divide," *New York Times*, December 3, 2011.

7. Ibid.

8. Ibid.

9. "President Obama Details Plan to Win the Future through Expanded Wireless Access," White House press release, February 10, 2011.

10. "FCC Reforms, Modernizes Lifeline Program for Low-Income Americans," Federal Communications Commission website, February 6, 2012.

11. "Original Obamaphone Lady," YouTube, September 26, 2012, www.youtube.com/watch?v=tpAOwJvTOio?.

12. "Mozilla Announces Global Expansion for Firefox OS," Mozilla press release, February 24, 2013.

13. Because the X phones are uncertified, they can't link to Google Play, but users can download apps from a special Nokia store, third-party app stores, and directly from developers. (Ian Delaney, "Xtraordinarily Xcellent: the Nokia X Family," Nokia Conversations website, February 24, 2014.)

14. Ibid.

15. Tom Warren, "This Is Nokia X: Android and Windows Phone Collide," The Verge, February 24, 2014.

16. Natasha Lomas, "Nokia Forks Android in Mobile Services Push," TechCrunch, February 23, 2014.

17. "Technology Leaders Launch Partnership to Make Internet Access Available to All," Facebook press release, August 21, 2013.

18. Eric Schmidt, "Preparing for the Big Mobile Revolution," *Harvard Business Review* website, 2011.

19. Ibid.

20. John Paczkowski, "Google's Schmidt: 'Our Goal with Android Is to Reach Everyone,' " All Things D, April 16, 2013.

21. Melanie Lee, "Google Controls Too Much of China's Smartphone Sector: Ministry," Reuters, March 5, 2013.

22. Grobart, "Tim Cook: The Complete Interview," *Bloomberg Businessweek*.

23. Javed Anwer, "Microsoft Offers Windows Phone OS Free to Indian players," *Times of India*, March 13, 2014.

24. "Microsoft to Acquire Nokia's Devices & Services Business, License Nokia's Patents and Mapping Services," Microsoft press release, September 3, 2013.

25. "An Open Letter from Steve Ballmer and Stephen Elop," Nokia press release, September 3, 2013.

26. 4Afrika website, Microsoft website.

27. Fernando De Sousa, "Connecting the Unconnected—Why We Are All 4Afrika," AllAfrica.com, July 9, 2013.

28. Matt Warman, "Huawei Plots to Overtake Apple and Samsung," *Telegraph*, March 2, 2013.

29. "Lenovo to Acquire Motorola Mobility from Google," Lenovo press release, January 29, 2014.

30. Steven Chase and Boyd Erman, "Lenovo Pursued BlackBerry Bid, but Ottawa Rejected Idea," *Globe and Mail*, November 4, 2013.

31. John Paczkowski, "Meet Xiaomi, the Biggest Smartphone Company You've Never Heard Of," All Things D, April 15, 2013.

32. "Nokia and Windows Global Momentum Continues," Kantar Worldpanel ComTech press release, April 11, 2013.

33. "Apple-China Mobile Deal a 'Watershed' Moment: Tim Cook," CNBC.com, January 15, 2014.

34. Jessica E. Lessin, "For App Makers, China Is Untapped and Untamed," *Wall Street Journal*, March 7, 2013.

35. Saritha Raijan, "Cost of Cool in India? An iPhone," *New York Times*, January 12, 2014.
36. "Developer Economics Q3 2013" report, VisionMobile, July 2013.
37. Ibid.
38. Ibid.
39. Miguel Helft, "Lenovo CEO on Apple, Samsung,' " *Fortune*, January 30, 2014.
40. Mat Smith, "Samsung CEO Promises to Deliver Devices with 'Folding Displays' in 2015," Engadget, November 5, 2013.
41. "Cortana (Yes!) and Many, Many Other Great Features Coming in Windows Phone 8.1," Microsoft Windows Phone blog, April 2, 2014.
42. Ibid.
43. Fairphone website.
44. Comment in response to "Our Role as a Social Enterprise in the Industry," Fairphone blog, June 20, 2013.
45. Comments in response to "Made with Care: Social Assessment Report," Fairphone blog, December 10, 2013.
46. Ibid.
47. Bibi Bleekemolen, "Taking Back Phones for a Circular Economy: E-waste in Ghana," Fairphone blog, November 21, 2013.
48. Comment in response to "Production and Distribution Update," Fairphone blog, December 30, 2013.
49. Ibid.
50. "Section 1201 Rulemaking: Fifth Triennial Proceeding to Determine Exemptions to the Prohibition on Circumvention," United States Copyright Office, October 2012.
51. "Mobile App Developers: Start with Security," Federal Trade Commission brochure, February 2013.

Index